**Living with the shore of Puget Sound
and the Georgia Strait**

Living with the Shore

Series editors: Orrin H. Pilkey, Jr. and William J. Neal

The Beaches Are Moving: The Drowning of America's Shoreline
New edition Wallace Kaufman and Orrin H. Pilkey, Jr.

From Currituck to Calabash: Living with North Carolina's
barrier islands *Second edition* Orrin H. Pilkey, Jr., et al.

Living with the South Carolina shore
William J. Neal et al.

Living with the East Florida shore
Orrin H. Pilkey, Jr., et al.

Living with the West Florida shore
Larry J. Doyle et al.

Living with the Alabama-Mississippi shore
Wayne F. Canis et al.

Living with the Louisiana shore
Joseph T. Kelley et al.

Living with the Texas shore
Robert A. Morton et al.

Living with Long Island's south shore
Larry McCormick et al.

Living with the California coast
Gary Griggs and Lauret Savoy et al.

Living with the New Jersey shore
Karl F. Nordstrom et al.

Living with the shore of Puget Sound
and the Georgia Strait

Thomas A. Terich

Sponsored by the National Audubon Society_{TM}

Duke University Press   Durham 1987

The National Audubon Society and its Mission

In the late 1800s, forward-thinking people became concerned over the slaughter of plumed birds for the millinery trade. They gathered together in groups to protest, calling themselves Audubon societies after the famous painter and naturalist John James Audubon. In 1905, thirty-five state Audubon groups incorporated as the National Association of Audubon Societies for the Protection of Wild Birds and Animals, since shortened to National Audubon Society. Now, with more than half a million members, five hundred chapters, ten regional offices, a twenty-five million dollar budget, and a staff of two hundred seventy-three, the Audubon Society is a powerful force for conservation, research, education, and action.

The Society's headquarters are in New York City; the legislative branch works out of an office on Capitol Hill in Washington, D.C. Ecology camps, environmental education centers, research stations, and eighty sanctuaries are strategically located around the country. The Society publishes a prize-winning magazine, *Audubon*, an ornithological journal, *American Birds*, a newspaper of environmental issues and society activities, *Audubon Action*, and a newsletter as part of the youth education program, *Audubon Adventures*.

The Society's mission is expressed by the Audubon Cause: to conserve plants and animals and their habitats, to further the wise use of land and water, to promote rational energy strategies, to protect life from pollution, and to seek solutions to global environmental problems.

National Audubon Society   950 Third Avenue   New York, New York 10022

© 1987 Duke University Press, all rights reserved
Printed in the United States of America on acid-free paper ∞
Library of Congress Cataloging in Publication Data appear on the last printed page of this book.

Publication of the various volumes in the Living with the Shore series has been greatly assisted by the following individuals and organizations: the American Conservation Association, the Charleston Natural History Society, the Coastal Zone Management Agency (NOAA), the Geraldine R. Dodge Foundation, the William H. Donner Foundation, Inc., the Federal Emergency Management Agency, The Fund for New Jersey, the George Gund Foundation, the Mobil Oil Corporation, Elizabeth O'Connor, the Sapelo Island Research Foundation, the Sea Grant programs in Florida, Mississippi/Alabama, New Jersey, New York, and North Carolina, an anonymous Texas foundation, M. Harvey Weil, and Patrick H. Welder, Jr. The Living with the Shore series is a product of the Duke University Program for the Study of Developed Shorelines, which is funded by the Donner Foundation.

Contents

Foreword

Figures and table

Figures

Table

Foreword

Puget Sound is different. What most people think of as a broad expanse of water is actually a collection of long, narrow straits and passages, sheltered bays, and quiet coves. Puget Sound's numerous islands also make it unique. These picturesque islands are mantled under a dense cover of Douglas fir, red cedar, and ferns, but a few are little more than shields of bare solid rock protruding above the water.

For many people the attraction of Puget Sound and Georgia Strait is their calm and seclusion. Calmness generally prevails; however, seclusion is becoming harder to find as more people seek permanent or seasonal homes around the shoreline. Puget Sound, like most coastal areas around the United States, is under tremendous development pressure. It is obvious that demand for coastal space will continue and perhaps accelerate as western Washington realizes its potential in Pacific Rim trade and commerce. What is not so obvious are the impacts these developments will have on the shoreline.

The coastal zone is fundamentally different from inland areas, primarily because of its inherent instability. The interface of land and sea is highly mobile. This mobility can cause significant problems for those who construct immobile structures along the capricious coastline. Obviously risks cannot be eliminated, but they can be minimized. Americans have had a great deal of experience with shoreline development elsewhere. The rush to the Puget Sound shore comes relatively late. We can benefit from the mistakes made elsewhere.

Puget Sound and its shoreline are a valuable resource and, like other natural resources, should be used for the benefit of all the people. Wise use requires some knowledge of the shoreline's physical characteristics and the forces that shape and change it. Armed with this kind of information, development decisions can be made maximizing the benefits of the coastline while minimizing the risks.

The purpose of this book is to provide an easily readable and understandable source of information about the shoreline of Puget Sound and Georgia Strait. The book can be used by the curious for general information or as a guide when buying or building around the sound. The information in this book will not substitute for a thorough on-site inspection of a particular property by a professional, but it will give some general clues and tips for recognizing some potential problems common to Puget Sound and its shoreline.

This book was produced through the efforts of many people. In spite of an extremely busy professional schedule, Peter H. F. Graber, attorney-at-law and recognized authority on coastal management law, prepared the chapter on coastal land-use planning and regulation. Several graduate students and coastal researchers willingly supplied research information extremely useful to the book. They include Dana Blankenship, Mike Chrzastowski, Brad Harp, David Hatfield, Ed Jacobsen, Ralph Keuler, and Bruce Taggart. A good bit of writing and editing was done by series coeditor Orrin H. Pilkey, Jr., who spent a couple of weeks in the summer of 1985 looking at the beautiful shoreline of Puget Sound. Much-needed editorial advice and constructive criticism came from my wife Maureen and our good friend Karen Leque. Finally, I would like to express my appreciation to the Western Foundation and Bureau of Faculty Research at Western Washington University, which provided some financial and valuable clerical support in the preparation of this book.

The present volume is one of a series being published by Duke University Press. The series will eventually cover all coastal states. The first volume, *From Currituck to Calabash: Living with North Carolina's Barrier Islands*, is concerned with the barrier island coast of North Carolina. The success of this book in promoting safe and sound use of the North Carolina islands led to support from federal agencies to produce the other books. Most of the state books are closely patterned after *From Currituck to Calabash*. Several sections, such as the philosophy of shoreline conservation, are repeated here with minor revisions. With the use of this book we hope to aid Puget Sound citizens in evaluating the safety and longevity of various portions of their shore. We don't want anyone to be in the frustrating and even tragic position of saying, "How was I to know that . . . ?"

As part of this coastal safety series Van Nostrand-Reinhold Company published *Coastal Design: A Guide for Builders, Planners, and Homeowners*, by Orrin H. Pilkey, Sr., Walter D. Pilkey, Orrin H. Pilkey, Jr., and William J. Neal, in 1983. *Coastal Design* emphasizes coastal construction principles to a much greater extent than the individual state books. We recommend that the prudent coastal citizen also obtain this reference.

The overall project of producing these books is an outgrowth of initial support from the National Oceanic and Atmospheric Administration through the Office of Coastal Zone Management. The project was administered initially through the North Carolina Sea Grant Program. Support from the Federal Emergency Management Agency allowed us to expand the book project to all coastal states. The technical conclusions presented herein are those of the authors and do not necessarily represent those of the supporting agencies.

The series' editors owe a debt of gratitude to many individuals for support, encouragement, and information for the series project. Peter Chenery of the North Carolina Science and Technology Research Center and Richard Foster of the Federal Coastal Zone Management Agency gave us encouragement and support at critical junctures of this project. Doris Schroeder has helped us in many ways as Jill-of-all-trades over more than a decade. Jane Bullock and Mike Robinson of the Federal Emergency Management Agency worked to help us chart a course through the shifting channels of the federal bureaucracy. Dennis Carroll, Jim Collins, Jet Battley, Peter Gibson, Melita Rodeck, Richard Krimm, Chris Makris, and many others also helped us through the Washington maze. Special thanks are extended to Tonya Clayton and Leslie Droege for drawing the line illustrations. Finally, we are in the debt of many coastal residents, fellow geologists, coastal engineers, and state and local government officials too numerous to name who enthusiastically provided us with a wealth of data, ideas, and "war stories."

We dedicate this work to all who have helped, and to all who come to enjoy the shoreline of Puget Sound.

> Thomas Terich, senior author,
> Peter H. F. Graber, contributor, and
> William J. Neal and Orrin H. Pilkey, Jr.,
> series editors
> October 1986

Living with the shore of Puget Sound and the Georgia Strait

Figure 1.1 This satellite image shows western Washington's dynamic land-scape. The Puget Lowland and Georgia Strait are bordered to the east and west by high mountains.

1 *A Puget Sound perspective*

How did it get here?

A clear day view from nearly any place in Puget Sound or Georgia Strait reveals the snowcapped peaks of the Olympic Peninsula to the west and Cascade Range to the east (fig. 1.1). Puget Sound occupies a huge trough between the two parallel mountain systems (fig. 1.2). Dynamic forces deep within the earth have shaped Puget Sound and its neighboring mountains. These forces remain active today, as witnessed by the eruption of Mount St. Helens and continuing earthquake activity throughout western Washington.

The veneer of the region has been sculpted by at least four separate episodes of glaciation over the past few million years. Great tongues of ice pushed down the Fraser River Canyon in British Columbia, following the Puget Lowland, advancing as far south as Bremerton (fig. 1.3). This massive ice lobe was often joined by smaller glaciers emerging from highland valleys of the Cascades and Olympic Mountains. In some areas the ice was over a mile thick. Like a huge bulldozer, it reshaped the land, excavating and depositing great volumes of soil and rock.

Most of the deep, narrow channels in Puget Sound were dug in the bedrock by the scouring action of the ice. In the San Juan Islands and the adjacent Canadian Gulf Islands, the elongate shape and orientation of the islands clearly show the effect of a north-south-moving body of ice. On a smaller scale, large scratches or striations are plainly visible on hard rock exposures along the shoreline, revealing the strength of the ice. Great masses of rock and soil debris were also carried by the glaciers and deposited throughout the Puget Lowland from Blaine to Olympia, creating the gently rolling landscape we see today.

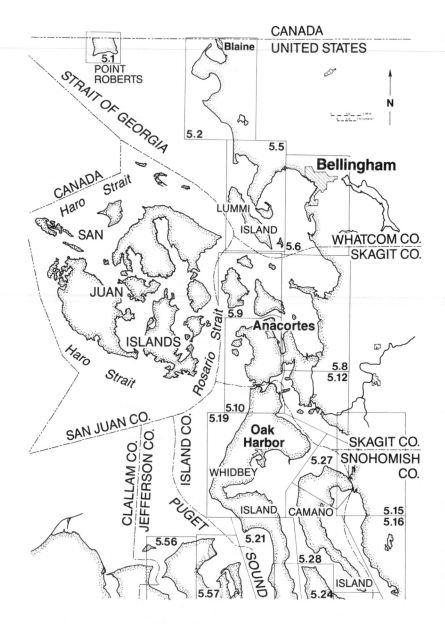

Figure 1.2 Index map of Puget Sound showing the main islands, inlets, and communities. Numbers shown refer to this book's site index maps.

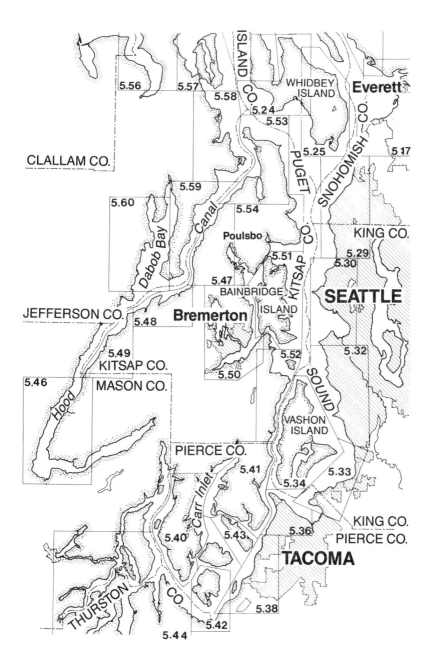

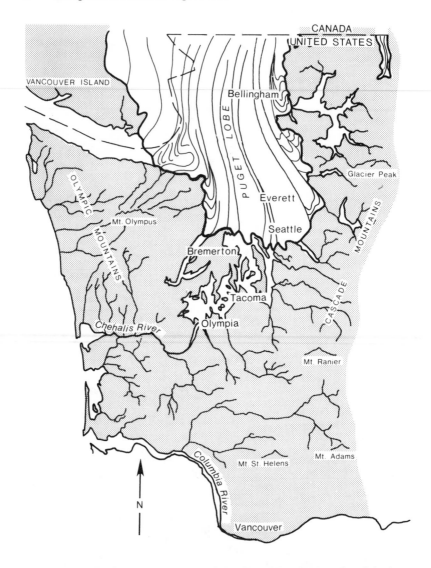

Figure 1.3 An artist's conception of the Puget Lowland as it might have looked over 20,000 years ago when glaciers covered the landscape. Meltwater flowed west from what is now Olympia into the Pacific Ocean; icebergs floated through the Straits of Juan de Fuca.

What the glaciers left behind

Our glacial legacy is clearly revealed by layers of sands and clays exposed in road cuts and shore bluffs around Puget Sound (fig. 1.4). The story of past glaciation can be easily reconstructed by an expert examining the exposed strata. Most commonly these sediments are mixtures of a bouldery clay called glacial till. When laid down under advancing ice, till becomes very dense and compact, forming a very hard but still unconsolidated and erodible layer. If deposited by retreating ice, till loosely blankets the landscape. Loose tills absorb a great amount of ground water, and saturation reduces its cohesiveness, making it very susceptible to sliding. Many of the landslides on shore bluffs around the Puget shoreline are caused by failures within glacial tills (fig. 1.5).

Near Olympia, at the southern end of Puget Sound, the land surface is mantled by rock and gravel, which is also of glacial origin. These deposits are called "outwash sands." They represent the residue that washed out of the southernmost reaches of the glaciers in the Puget lowlands. It is not uncommon to see thick layers of rounded rocks and gravel overlying a dull gray mass of clayey till. These two materials react very differently to ground water movement. The outwash sands and gravel allow water to move through freely, while the tills, because of their clay content, are generally a barrier to ground water. As a result, ground water will quickly pass down through the outwash materials and hit the top of the till layer, at which point the water is forced to flow

Figure 1.4 High sand and gravel bluffs like these at Point Defiance rim the shorelines of western Washington, revealing the glacial legacy of the region.

Figure 1.5 The glacial history of Puget Sound is shown in the shore bluffs as layers of differing soil and rock materials. When saturated by late winter and early spring rains, these bluffs readily slide onto the lower beaches. **Figure 1.6** Sea level has risen dramatically over the past few thousand years because of climatic warming and the melting of the ice caps.

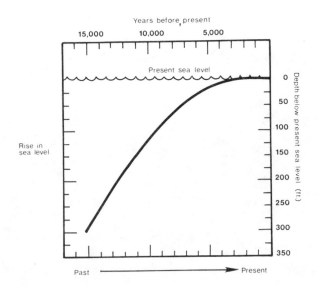

horizontally, eventually emerging out of the shore bluffs or road cuts as springs or seeps. Landsliding is very common under these conditions because the till layers act like slide surfaces. The periods of greatest landslide activity in Puget Sound tend to be in late winter or early spring when the ground is most saturated with water. Our glacial legacy has left us with a natural hazard that must be considered when developing shorefront property in Puget Sound.

Sea-level changes

Each retreat of the glaciers was followed by another catastrophic event, a sudden and rapid rise of sea level in Puget Sound. As the ice sheets worldwide melted back in response to climatic warming, tremendous volumes of meltwater were released. This resulted in a worldwide raising of sea level by as much as 300 feet (fig. 1.6). There is some geologic evidence indicating that, for a brief period of time, sea level in Puget Sound might have stood as much as 600 feet higher than it is today.

Apparently, the history of sea-level changes in Puget Sound has not always followed the history of open ocean sea levels. The differences might be accounted for by some local sinking of the land due to the weight of the ice on top of it, or to ice damming of the Strait of Juan de Fuca, causing a significantly higher water level in Puget Sound than in the open ocean.

As the great tongue of ice melted, huge icebergs laden with mud, sand, and gravel floated across the flooded Puget basin. The sediment rained down, covering much of northern Puget Sound with a mixture of glacial and marine sediments that geologists call "glaciomarine drift." This material blankets much of the area today. Glaciomarine drift is similar to glacial till, except that it is embedded with seashells and other remains of marine organisms.

Deciphering the history of the stands of sea level in Puget Sound is complicated. Suffice it to say that sea level here, as well as elsewhere in the world's oceans, has been relatively stable over the last 5,000 years. Thus, the general configuration of the shoreline in Puget Sound has been in place since that time. Wave and tidal forces have molded the shoreline we see today. These forces continue to operate, but now man has entered the scene. Developing the shoreline, we interrupt or try to alter wave and tidal forces to adjust to our needs instead of nature's.

But the sea-level rise story doesn't end with the demise of the last ice age. Recent evidence summarized by the Environmental Protection Agency indicates that the sea level is once again on the rise and that the rate of rise is accelerating. In fact, according to the EPA, the most likely rise by the year 2100 is between four and seven feet above present levels. The maximum expected change in elevation of the sea surface by 2100 is ten feet! The present rate of worldwide sea-level rise is one foot per century.

The presumed cause of the present-day sea-level rise is melting of the Antarctic ice sheet, which in turn is happening because of the "greenhouse effect." The greenhouse effect is produced by rapidly accumulating amounts of carbon dioxide in our atmosphere, contributed primarily by our burning massive amounts of coal and oil to run our society. Since it is highly unlikely that we will stop using coal and oil, the sea-level rise is very likely to continue.

Raising the level of the sea surface by amounts just discussed will have a profound effect on Puget Sound. Most important of all, erosion rates and flooding will significantly increase. In fact, one of the underlying reasons for erosion and flooding right now in Puget Sound is undoubtedly the sea-level rise.

2   Coastal dynamics

Coastal Puget Sound is mainly vertical. Homes are built upon high shore bluffs that look down through water that deepens rapidly within short distances from shore (fig. 2.1) The high bluffs and numerous islands that rise above that water are mostly a product of our dynamic geologic and glacial history. Since the leveling off of sea level about 5,000 years ago, waves and tidal currents have been gnawing away at the land. The silts and clays have been swept away by currents, while the coarser sands, pebbles, and cobbles have remained behind, evolving into the beaches we have today. Nearly all of the beaches along the shores of Puget Sound and Georgia Strait are a direct product of sea cliff erosion and are indirectly related to glaciation that shaped the Puget Sound landscape thousands and even millions of years earlier. Most of the pebbles and cobbles on the beaches of Puget Sound were carried there by the ice and embedded in the landscape to be eroded out by wave action forming the beaches we enjoy today (fig. 2.2).

The rate of land erosion in Puget Sound was probably much greater when sea level first stabilized about 5,000 years ago. As time progressed and erosion continued, the beach buffer between land and sea widened. Now, in many places around Puget Sound, waves only attack the bluffs when the water level is at its highest at the coincidence of high tides and storm waves (fig. 2.3).

Beaches—sources and sinks

Some years ago scientists at Scripps Institute of Oceanography produced a film entitled, "Beach: A River of Sand." The film clearly demonstrates how beaches are like moving rivers of sand flowing from "source" areas to places where they are lost, called "sinks." In the case of California, where the film was made, rivers are the source areas. Rivers bring great

Figure 2.1 (top) The high shore bluffs around the sound and straits provide spectacular views, but builders should use caution to reduce landslide hazards. **Figure 2.2** (middle) Most of the region's beaches are narrow and made of sands and gravels eroded from nearby shore bluffs. **Figure 2.3** (bottom) Waves most often gnaw at the shore bluffs at high tide and during winter storm surges.

quantities of sand to the coast where waves move it along the shore. Downcoast, the sand is lost to large offshore canyons, which brings an end to the beach until another river or source is found. Puget Sound beaches behave in a similar way; however, they have different "sources" and "sinks."

Most of the sediment composing Puget Sound beaches comes directly from nearby bluffs (fig. 2.4). This explains why most of the beaches in the sound are made of pebbles and cobbles. Where the shore bluffs are mainly sand, the adjacent beaches are also sandy. In Puget Sound the beach is not generally lost to deep offshore "sinks" as in the California example. Most of it comes to rest in sand bars, spits, and small capes in shallow water. These features represent an end product of shore bluff erosion. Each is an accumulation point of the sediment that has been deposited and molded into unique shapes by the action of waves and currents. These places are particularly attractive development sites for residential and recreational homes (fig. 2.5). They offer spectacular views, easy water access, and often have calm water bays on one side and the open sound on the other. These features can be fairly safe for development so long as elevation is high enough to avoid storm waves and flooding and so long as there remains an uninterrupted supply of beach sediment feeding them. When sediment supply is interrupted or cut off completely, these shore features erode rapidly and can even be severed from the mainland.

Waves and littoral drift

If there were no waves at sea or on Puget Sound, there would probably be no reason to write this book. Waves essentially represent energy moving through the water. That energy is ultimately expended upon the shoreline, working to erode, transport, and deposit beach sediment. Waves get their energy from the wind blowing over the water surface. The faster and the longer the wind blows, the higher the waves. Maximum wave height is also controlled by the size of the water body over which the wind is blowing. The larger the body of water, the bigger the wave that is developed. For example, a 30 mph wind blowing for 24 hours over the north Pacific can generate waves over 20 feet high. In smaller Puget Sound, those same wind conditions will create waves only 2 to 3 feet in height (fig. 2.6).

The wind translates its energy to the water surface, causing waves to develop. The waves travel through the water losing little energy. When the shore is reached, the wave crashes upon the shore, causing turbu-

Figure 2.4 (left) One person's erosion is another person's beach! As most beach sediment comes from shore bluff erosion, someone's property must erode to allow another's beach to grow.
Figure 2.5 (below, top) Long spits like Sandy Point in Whatcom County are popular building sites. Unfortunately, because they require a constant supply of sediment and are low in elevation, they are vulnerable to erosion and flooding. **Figure 2.6** (below, bottom) Wave energy both creates and destroys beaches. Placid summer surf can easily become threatening winter breakers.

lence. Beach sediment is stirred up. Two forces push the sediment along the shore: the back and forth (swash and backwash) action of the surf and a nearshore current called the longshore current. On a typical quiet day neither of these processes is well-developed along the beaches in Puget Sound. This is because waves are relatively small and the beaches are narrow and composed predominately of large-grained sediment so heavy that waves can't easily move the cobbles and pebbles. One has to wait for a storm to see the action on a sound beach. On Washington State's Pacific Ocean beaches, however, these forces can be seen and felt every day. The breaking waves move the sand, and the action of the surf drives it up the beach face. The water laden with sediment then flows back to sea in large arcs. Each successive swash drives the arclike motion of the sediment down the beach in the direction of wave approach. At the same time a current develops in the surf zone. This longshore current carries wave-disturbed sediment parallel to the shore (fig. 2.7). This "river of sand" may move along the shore for great distances. However, along Puget Sound beaches the "rivers of sand" may be as short as a few tens of yards where the sediment flow is interrupted by rocky headlands or other shore features. Several thousands of cubic yards of sediment can be carried along a beach by littoral drift during a single storm. The littoral drift of beach sediment in Puget Sound occurs mainly, but not exclusively, when waves are driven by storm winds. During these periods, waves have enough energy to move the pebbles and cobbles along the beach.

Most beaches in Puget Sound will have a preferred net direction of littoral sand transport when looked at from a one-year standpoint. This preferred direction of net transport of beach material is highly variable

Figure 2.7   Beaches are like rivers of sediment moving along the shore from places of origin to destination.

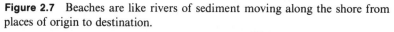

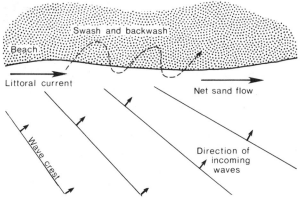

because the shorelines in this region are so irregular, because their orientations are so different, and because the open water distances over which waves are created are so variable. Generally speaking, where there are long stretches of north-south-trending shorelines, the net littoral sediment drift is driven to the north by strong storm waves coming from the south. Examples of relatively long north-south-trending shorelines with net northerly moving littoral sediments are the areas from the Nisqually Delta (near Olympia) north to Point Defiance (at Tacoma), and from Des Moines north to Seattle's Alki Point. Directions of net littoral transport for these and other stretches of shoreline are shown on the site analysis maps in chapter 5.

Beach erosion

Most Puget Sound and Georgia Strait shorelines are "eroding" or "retreating" as researchers call it. "Shoreline retreat" is the scientific term, but "beach erosion" is the more common term used by beachfront property owners. Natural rates of shoreline retreat are measured in inches per year, but such rates are highly variable. On hard rock shorelines, rates may be close to zero, but on high and sandy bluffs the retreat might be several feet in a single storm. There are many causes of shoreline retreat including sea-level rise, usual and unusual storms, man-made influences, or a combination of all three factors.

Beaches are a product of waves. Waves create them, and waves take them away. Under natural conditions most beaches erode during periods of storm wave activity. During storms the waves are larger, with more energy to move pebbles and cobbles along the shore. If the storm waves coincide with a high tide, they can do their work higher on the beach, causing maximum erosion. During the calmer periods of summer, waves exert little energy on the beaches (fig. 2.8). In fact, smaller, calmer waves can bring back some of the finer sands from offshore up on to the beaches, conveniently creating a softer beach surface for sunbathing! The main point to be emphasized here is that, in Puget Sound, shoreline erosion is almost exclusively a storm phenomenon.

It is important to make a distinction between beach and bluff erosion. Most of the beaches in Puget Sound are products of sediments eroded out of the shore bluffs. Thus, one man's eroding bluff creates another man's beach. Beaches, on the other hand, represent the residue of the erosion process. They might show some seasonal change in physical shape and sediment size, but as long as a supply of sediment is available, the beach should be similar in appearance over a long time. However, if the supply of sediment is interrupted or cut off, the beach will narrow

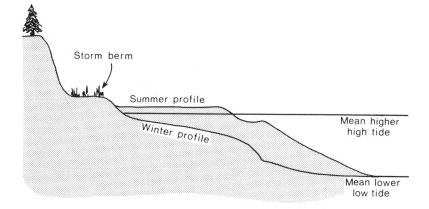

Figure 2.8 The profile of beaches changes with seasonal wave energy. They tend to be low and narrow in winter, wide and high in summer.

and retreat quickly. This disruption of sand supply is one of the chief causes of beach erosion in Puget Sound.

The sediment supply to beaches can be cut off or reduced by property owners who build seawalls in front of eroding shore bluffs or who dam the littoral drift with walls (called groins) built perpendicular to the shoreline. A cut-off beach sand supply translates into shoreline erosion. These topics will be discussed in greater detail in the next chapter.

Beach shapes

The beaches in Puget Sound collectively come in many shapes and sizes and are made up of a wide variety of materials (called sediment). Most beaches are narrow ribbons of loose sediment forming aprons around the shore bluffs (fig. 2.9). They are littoral conveyor belts moving sediment from source areas to sinks. At the sinks, the material is deposited, creating some of the major beach features commonly found in Puget Sound and discussed below.

Spits. A spit is an extension of a beach built out into deeper water but still connected to the land (fig. 2.10a). It is a sink at the receiving end of the littoral supply system. Spits commonly take the form of a long, narrow ridge of sediment extending across a bay. They tend to straighten out an otherwise sinuous shoreline. In some cases spits push out from the mainland like a large hook with multiple barbs on the inside of the hook. Ediz Hook, a classic spit found at Port Angeles in the Strait of Juan de Fuca, takes its name from its shape. A few miles to the west of

Figure 2.9   On the west shore of Whidbey Island wide, sandy beaches face the Straits of Juan de Fuca.

Ediz Hook is Dungeness Spit, a spit of huge proportions (nearly 8 miles long), with a large bay-side barb. Most spits in Puget Sound are small and are usually found extending across bay mouths, for example, Travis and Gibson spits across the mouth of Sequim Bay, a dozen miles west of Dungeness Spit.

As a rule, a spit will always grow or be pointing in the direction of net littoral drift. They are therefore good indicators of the local net direction of sediment movement.

Bars. These are underwater ridges of sand that parallel the shoreline and are seldom seen except at very low tides. These are important because during storms the big waves often "trip" over the·bars rather than smashing directly into the beach. In some conditions, a spit might grow completely across a bay mouth, in which case it is known as a "bay-mouth bar" (fig. 2.10b).

Tombolo. A term of Italian origin describing an above-water ridge of sediment connecting a nearshore island with the mainland. These are fairly common in this area; for example, the bar connecting Decatur Island and Decatur Head (fig. 2.10c).

Looped bar. A type of spit that has been shaped by wave action into a loop. Looped bars will often enclose a small bay or low marsh. An example of a looped bar in Washington can be seen on Whidbey Island, two miles south of Partridge Point (fig. 2.10d).

Cuspate forelands (capes). These are distinctive triangular or cusp-shaped deposits that project from the land into the water. They can range in size from less than one acre to several acres. In most cases they are formed where littoral drift currents from two different directions converge (fig. 2.10e).

Figure 2.10 A variety of low-lying depositional beach forms are found in the region. From top to bottom: (a) a small spit along the Hood Canal shoreline, (b) a baymouth bar, an underwater bar that has grown across the mouth of the bay, (c) a tombolo—a narrow ribbon of deposited sand and gravel—connecting offshore rocks and islands to the mainland, (d) a looped bar at Whidbey Island, (e) cuspate forelands, so named because these beach deposits take the shape of cusps.

Sliding shore bluffs

While gravity has some good points, it also has its drawbacks. One of the bad points is the force it exerts in lowering Puget Sound shore bluffs down to sea level. Gravity, with the help of waves, water, and man, is largely responsible for most of the landslide activity in western Washington. Few winters pass without local news reports of landslides threatening or destroying homes in the region. A combination of geology, climate, and oceanography make Puget Sound shore bluffs particularly vulnerable to sliding (fig. 2.11). The glacial makeup of the shore bluffs encourages instability. Stable when dry, the shore bluffs become particularly susceptible to sliding when wet. Thus, it is not surprising that many local slides occur during or just after periods of heavy rainfall. Studies have shown that the frequency of landsliding increases dramatically from January through March, coinciding with long periods of rainfall in western Washington. Watering bluff-top lawns during summer months also probably increases the landslide hazard in some cases.

Waves have a tremendous potential for causing coastal landslides. Waves attack the toe of a bluff, removing foundation support and causing slides. Surprisingly, except in a few isolated cases, waves are generally not the primary cause of shore bluff landsliding in Puget Sound. In most cases, waves simply wash away what has already fallen onto the beach due to other causes.

Man and his coastal developments play a major role in destabilizing

Figure 2.11 Gravity and ground water combine to weaken shore bluffs, resulting in frequent and active sliding.

shore bluffs. Diversion of excess water onto the slopes from roofs, paved areas, sewers and ditches, hillside excavations, road cuts, and fills leads to instability. Finally, indiscriminate cutting and clearing of natural vegetation on the bluff top and slope aid the erosion process. Developers, recognizing the vulnerability of the local landscape to sliding, should take great care in preparing shorefront building sites.

Is my slope sliding?

A slow walk and careful observation of telltale signs can provide some insights into the stability of your shorefront property. Some of the things to look for are:

Cracks or loose debris. Inspect the face of the slope or debris at the bottom of the slope indicating recent movement. Is there cracking in and above the slope revealing signs of tension and future movement?

Running water. Is there water running down the slope face or are there places where it might have flowed during the wet season? Is there water weeping out of the bluff face?

Vegetation. Is vegetation covering the slope or has it been removed? If large trees are not standing upright, this is a sure sign of active movement. If you must cut trees to improve the view, leave stumps and roots in the ground.

Bluff undercutting. This can occur due to frequent wave attack. Is there evidence, such as notching or a pile of drift logs at the base of the bluff, indicating storm wave attack?

Shore bluff stability is not always easy to detect. What looks quite stable can become very unstable under certain conditions. It is important to assume that slumping and sliding is likely to occur; with that in mind, keep as wide a space as possible between your home and the slope.

Coastal flooding

Having Puget Sound as a neighbor is a pleasant experience, but it's one neighbor you don't invite into your home! Coastal flooding is not a major regional problem around Puget Sound because most areas are elevated well above the water. It is a problem in low coastal areas where conditions for flooding are favorable. Low coastal areas, particularly those adjacent to rivers, as well as low-lying spits, bars, and cusps are vulnerable.

Water level in Puget Sound regularly rises and falls with the tides, which are predictable, but coastal flooding is an unusual event driven

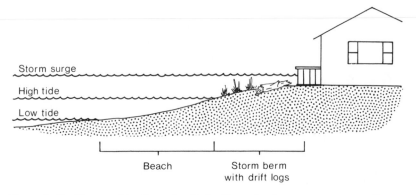

Storm surge

High tide

Low tide

Beach   Storm berm
                 with drift logs

Figure 2.12 The combination of a high tide and winter storm waves often leads to flooding of south-facing coastal areas of low elevation. **Figure 2.13** Drift logs pose an additional threat to beachfront homes. A late spring storm tossed logs and driftwood up to and into homes at Sandy Point facing Georgia Strait.

by forces superimposed on the tides. A high tide coinciding with the passage of a strong low-pressure area over Puget Sound is the most common condition that leads to coastal flooding (fig. 2.12).

The storm drives large surface waves and fast currents. When they run into the coast, the water "piles up" causing what is called a "storm surge." The amount of storm surging depends on the velocity, direction, and length of open water over which the wind is blowing. Low coastal sites and bay heads in the direct path of the wind and surging water are most vulnerable. Because most of the storm winds in Puget Sound are southerly, south-facing shores, particularly at bay heads, are likely candidates for flooding. Common areas of flooding in Puget Sound include Case and Carr inlets, Dabob Bay, and Port Susan. Large storm waves add to the inland reach of the flooding. Furthermore, the waves often carry drift logs that slam into shorefront homes (fig. 2.13).

If you are interested in buying some low (less than 10 feet above mean high tide) shorefront property and are concerned about the flooding potential, take advantage of the several sources of information available. Ask neighbors if they know of any past flooding. Better yet, talk with the town or county planners or zoning administrators. They can be very helpful and usually have access to flood maps produced by the Federal Emergency Management Agency (FEMA). These maps detail the locations and frequencies of river and coastal floods in western Washington and Puget Sound. More discussion of FEMA and flood insurance is given in chapter 6.

3 Man and the shoreline

Trying to stabilize the unstable

The fundamental problem of developing along the shoreline is that we build rigid immobile structures in a highly mobile environment. When the shoreline begins to change, we try to prevent it from doing so—and this has come to be known as coastal engineering or shoreline stabilization (fig. 3.1). Methods of stabilizing beaches and coasts range from the simple planting of dune grass to the costly and complex emplacement of huge seawalls and jetties. There are some very good reasons for trying to stabilize the coast; for example, navigation safety, port development, and storm protection of urban areas. Most of the large-scale coastal navigation and development projects on the West Coast are necessary and successful; some are not. All are costly and require constant maintenance.

There is another side to the coin, however. Most shoreline stabilization in Puget Sound is not carried out to protect navigation or ports. In fact, it can be argued that most Puget Sound seawalls are entirely unessential as far as serving any purpose for the common good. Instead, the miles and miles of walls and piles of rock simply serve to protect the private property of a relatively few individuals who built too close to the shoreline to begin with. Unfortunately, efforts to stabilize one part of the coast can lead to destabilization elsewhere. This has occurred over and over. Severe erosion often occurs "downstream" from a major coastal engineering project such as jetties. This same process operates on small-scale projects as well. The actions of one property owner to protect his or her coastal land can cause harmful "downstream" results.

Man has occupied and used the coast since time began and will continue to do so. The key to coastal use is the prevention and mitigation of potential problems. Coastal engineering as a science is advanced far enough to predict and prevent many problems dealing with coastal development projects. However, the average citizen, who knows little or nothing about coastal processes and engineering, should adopt a hazard prevention policy. This policy is best accomplished by a "safe" construction setback, thereby accommodating, at least temporarily, coastal changes with little or no threat to nearby structures. Where a "safe" setback is not possible or construction is after the fact, other options of erosion protection are available. They range from nonstructural to extensive structural measures. Several of these are presented here with a brief discussion of the advantages and disadvantages of each.

Construction setback

The most obvious way to avoid a hazard is to stay away from it! So it is with an eroding shoreline or unstable shore bluff. The prudent planner, recognizing that over the years some erosion or sliding is likely to occur, will build well back from the shoreline or shore bluff.

How far back is a "safe" building setback? That question is difficult to answer and will vary from place to place. When in doubt, ask questions of neighbors, longtime inhabitants, and town officials. Cities and counties bordering the marine waters of Washington State have shoreline construction setback requirements built into their shoreline management

Figure 3.1 Homes lining the shore are "protected" behind a fortress of seawalls and bulkheads, an unfortunate response to the threat of beach erosion and coastal flooding.

programs. These are available from city hall and the county courthouse. These requirements should be consulted prior to construction.

Advantages of construction setbacks:

- Reduces the threat to buildings
- Allows natural shoreline processes to operate without interference
- No impacts are suffered by neighboring properties
- Preserves the recreational and aesthetic values of a beach
- No long-term maintenance costs
- No permit problems

Disadvantages:

- Does not stop erosion or landsliding
- Might reduce water views
- Lot must be deep enough for a suitable setback

Moveback

In many cases existing shorefront homes are threatened by erosion. The costs and benefits of moving the structure back from the shore must be weighed along with other alternatives. Depending on the nature of the problem, a moveback can compare favorably to other alternatives and prove to be economically and aesthetically better in the long run. The advantages and disadvantages of a moveback are similar to the construction setback; however, there are a few additional items:

Advantages:

- Accomplished by contract with a professional mover
- One-time-only cost

Disadvantages:

- High cost
- Site must be deep enough to allow suitable moveback, or alternative site must be purchased

Do nothing

The do-nothing alternative involves total prohibition by local government of the use of any procedures or structures to prevent or slow down shoreline erosion. Thus individuals who own buildings adjacent to a moving shoreline must eventually move back or demolish beachfront buildings.

Advantages:

- No cost to taxpayer
- Preserves the recreational aesthetic values of a beach even in a very long-term sense
- Will greatly reduce future imprudent development

Disadvantages:

- High cost to individual property owner

Vegetation stabilization

Vegetation helps to shelter and bind soil and sand from erosion. Some varieties of plants are better than others for bank and beach stabilization. A local nursery operator, county extension agent, or soil conservation service office should be consulted. Bank stabilization in Puget Sound is best achieved by successive plantings, first with grasses or ivy, then with shrubs and trees.

Advantages of vegetation stabilization:

- Shelters and binds the soil or sand
- Reduces soil movement and rain erosion
- Self-maintained and renewable protection
- Aesthetically pleasing

Disadvantages:

- Will not stop severe erosion or sliding
- Vegetation might reduce views
- Vegetation might reduce access

Beach replenishment (nourishment)

Another alternative to help preserve an eroding beach is to restore it by adding beach sediment similar to the natural supply. This is not always practical or possible, but it is a good technique. In Puget Sound two important factors must be considered: cost and access to the beach. If this alternative is practical, it is encouraged because this is the only form of shoreline stabilization that improves, rather than degrades, the beach.

Beach replenishment has not been used to any large extent in Puget Sound, probably because there is no widespread public recognition of a need to maintain beaches. The usual immediate response to an eroding beach and threatened home is seawall construction. Yet supplies of sand

Figure 3.2 Concrete seawalls and wooden bulkhead stand as a line of defense against the pending assault.

and gravel suitable for dumping or pumping on beaches abound, thanks to the glaciers. The cost of replenishing open ocean beaches by pumping sand from offshore is usually a minimum of 1 to 1.5 million dollars per mile of shoreline. The cost of replenishing a Puget Sound beach is probably a small fraction of that. Furthermore, every replenishment job doesn't have to be a major one. Ten dump truck loads per beach house, placed every five years or so, would probably do wonders for many Puget Sound shorelines.

Advantages of beach replenishment:

- Most effective method of dissipating wave energy
- Beach is attractive and restored
- Restored beach will offer additional storm protection

Disadvantages:

- Beach will likely erode again, requiring additional material
- Cost and accessibility of "borrow" material might be prohibitive
- Temporary loss of beach fauna and flora

Bulkheads and seawalls

These two structural techniques are similar. They are vertical walls of concrete or wood constructed parallel to a beach, usually at the base of a bluff (fig. 3.2). These are the most common kinds of erosion structures found around Puget Sound because they rebuff wave attack and retain

Figure 3.3 Seawalls often lose in the battle with the sea. Failures are common because builders underestimate wave power.

the land behind. Unfortunately, seawalls and bulkheads often are not well designed or constructed, resulting in costly mistakes (fig. 3.3).

These stabilization structures are subject to numerous hydraulic forces that must be accounted for in their design and materials. Not only must they be constructed high enough to prevent storm wave overtopping, but they also must be implanted deep enough in the beach to prevent undercutting (fig. 3.4). In general, the maximum depth of expected scour is roughly equal to the highest breaking wave at the site; thus, maximum storm waves of 4 feet prescribe a footing depth of at least 4 feet. Some scour will still occur, and placing large rocks at the foot of the structure will help reduce this effect.

A seawall or bulkhead protects only the land behind it. The ends should be joined to neighboring structures if possible. Where none exist, wing walls or tie-ins to the adjacent land must be built to prevent wave flanking. Accumulating water and soil pressures will build behind bulkheads and seawalls. Drainage must be provided to allow water to escape from the landward side of the wall. This can be done by backfilling with gravel and having frequent openings (weep holes) along the lower part of the wall (fig. 3.5).

Additional strength can be gained through the use of tie-backs. Tie-backs anchor the upper part of the wall with steel cables to logs or other anchors deeply embedded into the beach or bluff.

This very brief explanation of seawalls and bulkheads is not intended to give design advice, but to illustrate some of the considerations. A professional should be consulted prior to any construction.

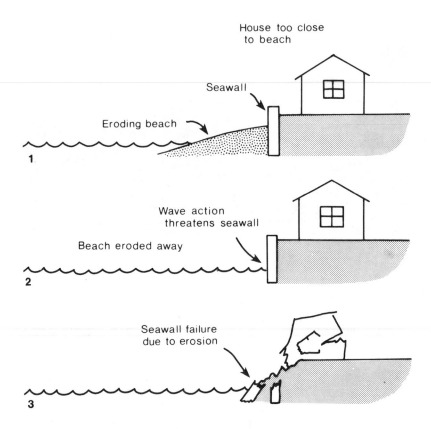

Figure 3.4 The typical progression of problems that can occur following the installation of a seawall.

Advantages of seawalls or bulkheads:

– Shields the land from wave attack
– Low maintenance if properly constructed

Disadvantages:

– Complex engineering design
– Depending on design and materials, can be costly (up to hundreds of dollars per foot)
– If improperly designed, requires regular, costly maintenance
– Special construction equipment required
– Limits access to beach
– Often aesthetically unpleasing

– Usually causes long-range, severe degradation (that is, narrowing or loss) of recreational beaches

Revetments (*riprap*)

Revetments are among the least expensive type of erosion protection. They consist of an armor facing of rocks, cement blocks, or other land material placed like a blanket on the sloping shore (fig. 3.6). Revetments are designed to absorb wave energy and reduce erosion by wave action.

Advantages of revetments:

– Relatively inexpensive
– Relative ease of maintenance
– Reduces wave erosion

Disadvantages:

– Protects against wave erosion only
– Frequent maintenance
– Reduces or destroys scenic and recreational qualities of beach
– Reduces access to beach

Figure 3.5   Seawalls are subjected to many hydraulic and mechanical forces.

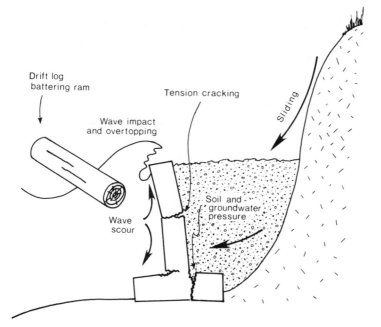

Figure 3.6 Large rocks piled on the shore to absorb wave attack are called riprap. Note fallen seawall facedown on the beach.

Groins

Groins are low walls built perpendicular to the shoreline. They are used primarily to trap sediment flowing in the longshore current, thus rebuilding an eroding beach. They are also useful in retention of sediment already on the beach. They are often, but not always, used in conjunction with seawalls. The problem with groins is that they trap sediment on one side and intensify erosion on the other, depending on the net littoral drift direction. Downdrift beaches become "starved" of sediment. Many lawsuits have been initiated by property owners claiming increased erosion damage resulting from nearby groin construction (fig. 3.7a, 3.7b).

Groins should be used in places of severe erosion only after careful study. A beach replenishment program is commonly used with groin construction to rebuild and preserve the beach. Groins are not usually necessary for erosion abatement in Puget Sound, and their use, except in very special circumstances, is ill-advised (fig. 3.8).

Advantages of groins:

– Quick trap for impounding beach sediment
– Retention of beach sand already present

Disadvantages:

– Leads to severe downdrift erosion
– Obstructs foot and vehicle traffic on beach
– Unsuitable in areas of low littoral sediment transport
– Can result in damage suits initiated by downdrift property owners

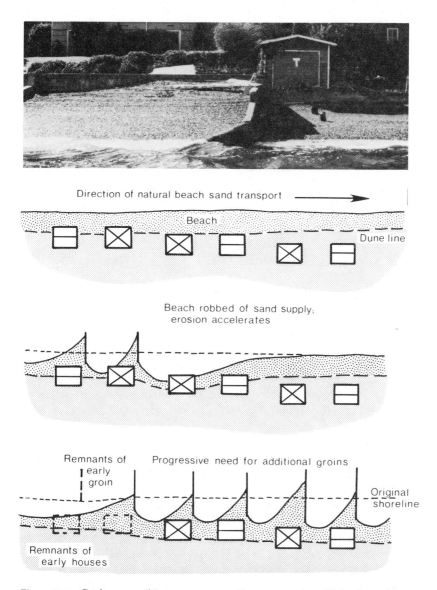

Figure 3.7 Groins are solid structures, usually concrete, installed perpendicular to the shore to trap moving longshore sediment. They act like dams, building beach sediment on one side and losing it on the other. (a) A photo of a groin in place on Days Island. Note the buildup of sediment on the left and the lack of sediment on the right. (b) A pictorial cartoon showing the sediment transport process and the effect of groins. Note that this leads to additional protection downdrift of the initial groins.

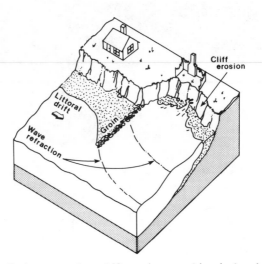

Figure 3.8 Groins cause downdrift erosion, resulting in beach loss and undermining of shore bluffs.

It should be obvious to the reader that the shore is a dynamic environment and a basic incompatibility emerges with the building of rigid erosion protection structures in this "fluid" environment. The installation of any erosion control or retaining structure should be made only after careful consideration of all nonstructural alternatives and with the advice of an expert. The emplacement of many structural controls, such as seawalls, can be an irreversible act with limited benefits. Often each structure must eventually be replaced with a bigger, more expensive one (table 3.1).

In addition to consulting an expert, a property owner should discuss his problems and plans with the local planning or building department. They are usually familiar with local conditions and are in a position to provide advice on alternative actions that are permitted.

In conclusion, we highly recommend that structural solutions to erosion problems be avoided if at all possible. Instead, we suggest that prudent setback be the primary means of combating Puget Sound erosion. If this alternative is not feasible, replenishment of eroding beaches with sediment from nearby sources should be undertaken.

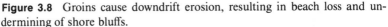

Table 3.1 Summary of erosion abatement techniques

Construc-tion setback	Vegetation	Beach nourishment	Bulkhead/seawall	Revetment	Groin
ADVANTAGES					
Reduces threat of destruction	Shelters and binds soil and substrate	Often makes beach more suitable for use	Shields the land from wave attack	Least expensive	Quick trap of beach sediment
Allows natural shore process to operate	Reduces soil creep and rain/gully wash	Does not affect downstream beaches	Low maintenance	Individual units allow settlement and replacement	Builds beach on updrift side
No impact on beach life	No impact on beach life				
DISADVANTAGES					
Does not stop erosion	Does not halt erosion	Does not stop erosion	Limits access and recreational use of beach and scenic view	Subject to settling	Downdrift beaches erode
Area must be available for relocation	Vegetation may reduce views	Cost and accessibility of borrow material varies	Complex design	Underlying material lost through joints	Unsuitable in low littoral transport areas
May reduce views	Reduces access	Affects beach life	Special equipment needed	Affects beach life	Legal problems may result
Special moving skills			Subject to failure if improperly designed	Limits recreational use of beach and scenic value	Affects beach life
			Affects beach life		Limits recreational use of beach
COSTS					
$3.00 to $5.00 per sq. ft.	Minor	$2.50 to $3.00 per cu. yd.	wood: $30 to $50 per lin. ft. concrete: $60 to $100 per lin. ft.	$30 to $80 per lin. ft.	wood: $10 to $25 per lin. ft. concrete: $60 to $100 per lin. ft.

A philosophy of shoreline conservation:
"We have met the enemy and he is us!"

In 1801 Postmaster Ellis Hughes of Cape May, New Jersey, placed the following advertisement in the Philadelphia *Aurora*:

> The subscriber has prepared himself for entertaining company who uses sea bathing and he is accommodated with extensive house room with fish, oysters, crabs, and good liquors. Care will be taken of gentlemen's horses. Carriages may be driven along the margin of the ocean for miles and the wheels will scarcely make an impression upon the sand. The slope of the shore is so regular that persons may wade a great distance. It is the most delightful spot that citizens can go in the hot season.

This was the first beach advertisement in America and sparked the beginning of the American rush to the shore.

In the next 75 years six presidents of the United States vacationed at Cape May. At the time of the Civil War it was certainly the country's most prestigious beach resort. The resort's prestige continued into the twentieth century. In 1908 Henry Ford raced his newest-model cars on Cape May beaches.

Today, Cape May is no longer found on anyone's list of great beach resorts. The problem is not that the resort is too old-fashioned, but that no beach remains on the cape.

The following excerpts are quoted from a grant application to the federal government from Cape May City. It was written by city officials in an attempt to obtain funds to build groins to "save the beaches." Though it is possible that its pessimistic tone was exaggerated to enhance the chances of receiving funds, its point was clear:

> Our community is nearly financially insolvent. The economic consequences for beach erosion are depriving all our people of much needed municipal services. . . . The residents of one area of town, Frog Hollow, live in constant fear. The Frog Hollow area is a 12 block segment of the town which becomes submerged when the tide is merely 1 to 2 feet above normal. The principal reason is that there is no beach fronting on this area. . . . Maps show blocks that have been lost, a boardwalk that has been lost. . . . The stone wall, one mile long, which we erected along the beach front only five years ago has already begun to crumble from the pounding of the waves since there is little or no beach. . . . We have finally reached a point where we no longer have beaches to erode.

From Cape May to Puget Sound

Along the East and Gulf coasts, there is widespread recognition among coastal dwellers that shoreline stabilization to halt erosion sooner or later ends up destroying beaches. Governor Bob Graham of Florida recently stated that the present generation does not have the right to destroy the next generation's beaches. It is not, however, a problem exclusive to only one side of our continent. In fact, southern Californians are slowly but surely coming to grips with the accelerating problem of disappearing Pacific beaches. As it turns out, the Atlantic and Pacific stories are identical; only the names have changed. Walls of various kinds cause loss of the beach either by the direct interaction of waves and structures forming currents that move sand away, or by the prevention of erosion of bluffs, thus cutting off the supply of fresh beach sand.

A beach management milestone of sorts was recently achieved in Monterey, California. There, the city officials refused to permit an apartment building owner to construct a wall to prevent his building from falling into the sea. Their rationale was that the public beach was more important than an apartment building and that construction of the wall would have destroyed the beach. Unless the courts reverse their decision, an apartment building will soon topple into the Pacific.

In North Carolina things have gone even further than that. The state passed a regulation in 1985 outlawing *any* form of hard stabilization. The spirit of the law is that temporary stabilization by sandbags is allowable, but only to buy time to move or demolish a threatened structure. The mettle of North Carolina's regulations will receive its real test when the first 10-story condo is about to fall into the sea.

All around the country, the various coastal states are finally beginning to view beaches as the national treasures they truly are. There is no question that the time will come when our remaining beaches will be like national parks, protected forever for the benefit of the public at large. They won't be like Yellowstone Park, however, where one can be sure that Old Faithful will be in the same location 50 years from now. Instead, our beach management policy will have to take into account that the sea level is rising and that the beaches are moving. Thus the beaches of the future may become mobile national parks!

Is there a beach degradation problem in Puget Sound? The answer is decidedly "yes," but there is little public understanding or even recognition of this. Miles and miles of Puget Sound shoreline are backed by rigid concrete walls or carefully stacked piles of rock that protect buildings, roads, and railroads. In front of these walls, beaches are narrowing with time. Lack of a high-tide beach is a common phenomenon. Sometimes the beaches have retreated so far, or the walls were built out so far, that there no longer is a beach at low tide! Perhaps the closest thing to a "New Jerseyized" beach in Puget Sound is Birch Bay, where walls abound and the beach, in places, looks more like a construction site than a national treasure.

So why isn't the public up in arms about the loss of their beach? Why haven't the city leaders of Seattle followed the path blazed by Monterey, allowing buildings to fall in, rather than permitting beach-destroying seawalls to be built?

Perhaps the answer lies in the fact that most Puget Sound beaches are not used by large numbers of swimmers. Every Puget Sound dweller knows that swimming along most (but not all) sound shorelines is indeed an experience to test one's character and tolerance to cold temperatures. Furthermore, many beaches are mantled with barnacle-covered rocks that don't make swimming a very pleasant experience.

Are the beaches of Puget Sound of any use to anybody? Of course they are. Just as in Monterey, California, or at Cape Hatteras, North Carolina, the beaches are places where we come to think, to brood, to smell the salty air, and to stare at a limitless and restless sea. We don't often see fishermen or surfers on Puget Sound beaches, but we see lots of strollers and clammers and, when the sun peeks through, one can even see legions of sunbathers.

Thus, the beaches of Puget Sound are being used. They are a precious state and national resource. Unfortunately, they are being degraded in many areas by unrestricted construction of erosion structures. Puget Sound beaches need the help of Puget Sound dwellers if they are to be preserved for our children, grandchildren, and great-grandchildren.

Truths of the shoreline

Certain generalizations or "truths" about the shoreline can be drawn from the American experience of shorelines on both sides of our continent. These truths are equally evident to both scientists who study the shoreline and to observant coastal residents. As aids to safe and aesthetically pleasing shoreline development, they should be the fundamental basis of planning and development along the shoreline.

There is no erosion "problem" until a structure is built on a shoreline. Beach erosion is a common, expected event, not a natural disaster. Much of Puget Sound's shore is in a state of change that includes erosion. However, shoreline erosion in its natural state is not a threat to the coast. It is, in fact, an integral part of coastal evolution and the entire dynamic system. When a beach retreats, that does not mean that it is disappearing; it is migrating. Whether the beach is growing or shrinking does not concern the visiting swimmer or stroller. It is when man builds a "permanent" structure in this zone of change that a problem develops.

Shoreline construction causes changes. The beach exists in a delicate balance with sediment supply, beach shape, wave energy, and sea level. Most construction on or near the shoreline changes this balance and reduces the natural flexibility of the beach. The result is change (that is, erosion) that often threatens the very structures that caused the problems to begin with.

Coastal erosion protection structures protect the interests of a very few, often at a high cost. Coastal erosion protection projects are almost always carried out to save beach property, not the beach itself. They are often shortsighted in terms of their design and impacts to neighboring beaches. The net result is damage to the beach, high maintenance costs, and sometimes lawsuits from neighboring property owners. The buildings and property that are protected are owned by a very small number of people, especially relative to the number of people who use the beach for recreation.

Shoreline erosion protection projects are irreversible. Because erosion protection structures often cause beach reduction, they must be maintained indefinitely. That is, the beach changes caused by seawalls make it even more necessary for the wall to remain in place to protect property. Furthermore, one structure often leads to a "domino effect," necessitating more wall construction by neighbors.

The solutions

1. Design to live with the changing coastal environment. Don't fight natural change with an artificial line of defense.
2. Consider all man-made structures near the shoreline as temporary.
3. Only as a last resort accept any engineered scheme for beach protection or preservation.
4. Base decisions affecting coastal development on the welfare of the public rather than the interests of a minority of shorefront property owners.
5. Begin to develop a coastal planning policy of noninterference. Recognize erosion as part of a natural process and realize that some developments might fail.

6. Consider the beach to be inviolate. Build nothing that will interfere with beach processes or degrade beach quality.
7. Recognize that some areas are more hazardous because of erosion, flooding, or landsliding problems, and establish development policies for these areas.

4   Selecting a site on the shores of Puget Sound and Georgia Strait

Mount St. Helens graphically showed the people of Washington and other states that some natural hazards are unavoidable no matter where you live. We are reminded of the fact over and over again when we hear of earthquakes in the West, tornadoes in the Midwest, flooding in the East, and hurricanes on the Gulf coast. Every part of the United States has its "unique" natural hazards. Moving from Kansas to California, one simply exchanges tornadoes for earthquakes!

Living at the shoreline, however, one is not only exposed to the same regional hazards as everyone else, but also to the additional problems of coastal flooding, beach and bluff erosion, and landsliding. These hazards are usually smaller in geographic extent, sometimes affecting only one property. But if it is your property, that's a catastrophe!

It is worth pointing out another rather unusual aspect of coastal hazards. One can't really compare tornado hazards in Kansas with shoreline living hazards on Puget Sound. The Kansas wheat farmer really has no choice but to live where tornadoes occur. The beachfront dweller, on the other hand, certainly doesn't have to live right next to the beach. In fact, a safer, more inland site would probably be less expensive!

Fortunately, nature often provides some clues about coastal flooding, landsliding, and erosion. If you know how to read those clues, it can save a lot of grief later on.

Natural clues

In Puget Sound a number of environmental features provide some clues to the natural history and present conditions of an area or site. Critically observing each of these features can help assess the relative level of safety or risk associated with the development of a given site.

Realtors have a saying, "Under all is the land," meaning that fundamental to any development is the land or property to be developed. This saying is also true from the perspective taken in this book, namely, that the physical and environmental conditions of a shore site are extremely important to its successful development use. Caution is advised when considering the purchase of a developed or undeveloped shorefront property. A wise piece of advice is to consult a land use or environmental specialist for a physical evaluation of the site prior to making financial and legal commitments. The major natural clues that should be observed include land elevation, vegetation, ground and surface water, and soil and rock exposures.

Elevation

Elevation above potential storm surging and coastal flooding is easily achieved when developing on a shore bluff, but not so easily achieved on a low-lying beach. Most low beaches in Puget Sound are a product of waves and currents, so it is wise to assume a potential flooding problem. If logs and wood debris are scattered on the site well inland beyond the beach, you know that storms have played havoc with the site in the past and will likely do so again in the future. More accurate information is best obtained by visiting the local planning or zoning office and seeing if your property lies within flood zones shown on flood hazard boundary maps. If your property is subject to flooding, all is not lost. Ask questions about the frequency and potential depths of flooding. Building precautions (such as construction on pilings) can be taken, and flood insurance might also be available. The majority of the Puget Sound shoreline is not likely to be flooded during storms. However, the Juniper Beach development in Island County is an example of one exception to this general rule. Almost always the exceptions are spits and sand bars constructed by littoral currents in the recent past. The other important types of areas where flooding may be expected during storms are the various river deltas, all of which are flat and low-lying.

Vegetation

Vegetation is often treated as a nuisance in Puget Sound, particularly when it obstructs water view. However, natural vegetation can be a real asset and should be treated accordingly. The very presence of grasses and small shrubs high on the backshore of a beach suggest low erosion potential and infrequent salt water intrusion. On shore bluffs, natural vegetation plays an even more important role. It helps to bind the soil surface and reduce erosion. Mature vegetation may even reduce the potential for landsliding. Look for a thick vegetation cover on the slope. If it is not there, ask yourself why. A slope too steep to support vegetation may be the result of recent landsliding and will be subject to landslides and erosion in the future. Crooked or bent tree trunks or exposed roots are clues to sliding (fig. 4.1). Old, tall Douglas fir trees growing on a shore slope are good indicators of long-term stability. On the other hand, a dense thicket of small alder can suggest ground disturbance and recent movement. Patches of horsetail weeds growing on a slope are clues of an overabundance of ground water and possible slope instability due to water saturation.

Ground and surface water

It is unfortunate that most real estate and building activity occurs in the spring and summer when weather conditions are mild. This often does not give the prospective buyer a chance to see the site during the other half of the year under more unpleasant weather conditions. If possible, it is advisable to visit the site during wet winter conditions. Observe the natural drainage and note places with standing water. Then recognize that development of the site will change the drainage characteristics and create impermeable surfaces concentrating runoff and drainage. Where sewer connections are not available, a septic system is required. A potential development site must pass a percolation test. Be sure to have

Figure 4.1 Exposed tree roots, bent tree trunks, and leaning trees are indicators of active shore bluff retreat and erosion.

the property sale contingent upon the successful passage of a "perk test." Last but not least, if at all possible, visit your site during a storm and see what the waves do to the shoreline!

Soil and rock layers

Most of the bluffs along the shores of Puget Sound are made up of unconsolidated combinations of gravel, sand, and mud. This is usually material that was brought to the area and deposited by glaciers a few thousand years ago during the last ice age (fig. 4.2). Frequently this material was deposited by rivers flowing in front of the glaciers as they melted. Such unconsolidated material is always highly susceptible to erosion by waves, and rapid rates of erosion are the rule rather than the exception. Erosion at the base of sea bluffs helps to maintain a steep bluff face, which in turn increases the potential for landsliding. The landsliding hazard is also affected by the proportions of coarse material (sand and gravel) and fine material (clay or mud layers). The presence of impermeable mud layers often causes water to concentrate along these layers and the combination of water and slippery mud lubricates the landslide.

Figure 4.2 The loosely consolidated shore bluffs at Marysville show signs of recent erosion and sliding.

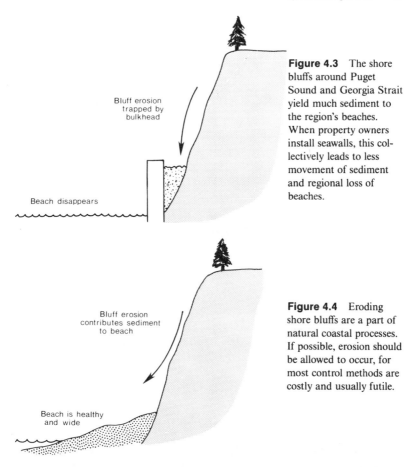

Figure 4.3 The shore bluffs around Puget Sound and Georgia Strait yield much sediment to the region's beaches. When property owners install seawalls, this collectively leads to less movement of sediment and regional loss of beaches.

Bluff erosion trapped by bulkhead

Beach disappears

Figure 4.4 Eroding shore bluffs are a part of natural coastal processes. If possible, erosion should be allowed to occur, for most control methods are costly and usually futile.

Bluff erosion contributes sediment to beach

Beach is healthy and wide

Naturally, people who live near eroding bluffs want to stop the erosion by construction of walls or whatever. The problem here is that the eroding bluff is also a major source of beach sand. Cutting off the supply of sand increases the rate of beach degradation in front of the wall and perhaps to neighboring properties as well (figs. 4.3 and 4.4).

A smaller percentage of Puget Sound shorelines are underlain by hard rock. For example, the shoreline south of Bellingham is largely made up of hard sandstones of the Chuckanut Formation (Cretaceous/ Tertiary age) (fig. 4.5). The southern half of nearby Lummi Island is made up of erosion-resistant greywackes and shales, as is much of the shore in the San Juan Islands. In general, such sites are environmentally far superior to glacial deposit shorelines from the standpoint of coastal erosion risks and landslide hazards.

Exposure

Exposure, or the amount of open water a shoreline faces, can be an important clue to the size of the waves that might be expected in a big storm. The greater the extent of open water, the larger the wave-making potential. Thus, other things being equal, one is better off building a beachfront house adjacent to a restricted body of water rather than on a shoreline facing a large area of open sea. For this reason alone, one can be assured that storm waves striking the west coast of Whidbey Island, which faces the open Strait of Juan de Fuca, have the potential to be much larger than storm waves striking the shores in the protected inlets around Olympia. Exposure can also increase flooding potential, especially at the north end of long, narrow, south-facing bays and inlets. Here, storm waves mounted upon a high tide drive water to extreme heights that flood normally dry coastal lands.

Shore types

It is impossible to pigeonhole or classify all of the shore forms found around Puget Sound and Georgia Strait, but there are at least six different major identifiable types of shorelines. Each is distinctive, yet each also shares certain characteristics with the other types. By defining

Figure 4.5   The rocky shores of the San Juan Islands and the Chuckanut shores near Bellingham are generally free of wave erosion and flooding. Nonetheless, caution must be taken against the threat of rock falls.

and having some understanding of each type, one can become an "environmental accountant," better prepared to evaluate the assets and liabilities of a site (fig. 4.6a, b, c, d, e).

Beach and high bluff

This type of shoreline is among the most common in Puget Sound (fig. 4.6a). It typically has a dry backshore beach of gravels and pebbles with some grass growth. Often, logs and wood debris are embedded in the sediment, tossed there by very high storm waves. Behind the backshore is a steep bluff varying from about 30 feet to 120 feet above sea level. The degree of beach and bluff stability is often illustrated by the width of the dry backshore and density of vegetation cover on the bluff slope. In more stable areas, the beach and dry backshore are relatively wide and the shore bluff is well vegetated. In areas of more active beach and bluff erosion, the backshore narrows or is nonexistent, and the bluff is lightly vegetated. However, remember that even heavily vegetated slopes in Puget Sound are susceptible to landsliding under certain conditions. Caution must always be taken: build well back from the edge of even apparently stable shore bluffs. The only bluffs that are "safe" are those made up of hard rock rather than the much more ubiquitous unconsolidated glacial material.

Beach and low shore bluff

Shore bluffs less than 10 feet high are also common to Puget Sound (fig. 4.6b). Low bluffs allow much easier beach access and are generally (not always) less susceptible to landsliding. Similar to high bluff shores, the degree of beach and low bluff stability is usually indicated by the relative width of the beach and density of bluff-face vegetation; the wider the beach and the more dense the vegetation, the safer the site. Shores with orientations toward frequent storms or large open bodies of water are most often exposed to storm waves and can erode quickly. Care in site selection should be taken accordingly. A wide setback from the edge of the bluff, whether high or low, is the best development policy.

Developed shoreline with seawall

Much of the Puget Sound shoreline is already developed with erosion protection structures, such as seawalls or bulkheads, in place (fig. 4.6c). Many people will take them for granted, but they should not. Assume a seawall was installed because of some erosion or flooding problem. Why

Figure 4.6 A wide variety of shore types are found in the region. Here are, from top to bottom, (a) a beach with high bluff, (b) a beach with a low bluff composed of glacial till, (c) a developed shore, (d) the delta/floodplain at Nooksack, (e) a low beach with no bluff at Dewey Beach.

else would it be there? Inquire as to the nature of the problem, how long the structure has been in place, if the structure is serving its purpose, whether the structure shows obvious signs of failure, and what is the likelihood of future costly repairs or reconstruction. It is wise to call in a specialist to evaluate the nature of the problem and the condition of the structure. Seawalls can aggravate an erosion problem, leading to a lowering and narrowing of the beach. Alternatives to seawall installations are advisable. More often than not, if you live behind a seawall, you can expect (1) the beach to gradually get narrower, (2) continuing maintenance costs, and (3) increasing potential for storm damage as the beach narrows.

Deltas and tidal flats

Deltas and tidal flats are usually environmentally sensitive areas in Puget Sound because they are important aquatic and bird habitats (fig. 4.6d). Flooding is perhaps the most common development hazard near these zones because of their low elevation. It is probably best that these areas be kept free of development except for obvious high ground. Nonetheless, check with local building or planning officials about flooding potential.

Rocky shores

Hard, rocky shores are quite unusual features save for a few regions of Puget Sound and Georgia Strait. Beaches are seldom associated with rock cliffs. Development in these areas is usually in niches and cliff tops, taking maximum advantage of views. Wave erosion is usually not a major problem because the hard rock rebuffs wave attack. However, rock falls can present problems. Some setback from the cliff edge is wise.

Beach and wide backshore (no bluff)

There are relatively few low beach development sites in Puget Sound. Most are sediment accumulation forms, "sinks" in the littoral drift system (spits, bars, forelands). Examples include Lane Spit on Lummi Island, Driftwood Shores in Island County, and Alki and Williams points in Seattle. These places are very attractive for residential development because they offer beach-level sites with easy water access (fig. 4.6e). Unfortunately, these areas are often very vulnerable to erosion and flooding. Verify any suspected flood problems with neighbors or local building officials. Check on the availability of flood insurance. Be particularly wary of flooding and erosion if the neighboring properties have seawalls. They did not go to the expense of installing seawalls for nothing! Furthermore, if some of the property owners also have installed groins to trap beach sand, it might be best to look elsewhere for shorefront property. Remember that these types of shores require a continuous and uninterrupted supply of sediment. If that sediment supply (littoral drift) is interrupted by a groin, severe downdrift erosion is the result.

The site: a checklist for safety

The following is a brief checklist for evaluating your shorefront homesite on Puget Sound. The questions are intended to present some of the obvious considerations essential to site safety. The list should not be used as a substitute for on-site inspection by a specialist, however. Its best use is in determining whether a specialist should be consulted for a more in-depth study of the site and for development recommendations.

1. Is the site on hard rock as opposed to unconsolidated or uncemented sands and gravels?
2. Is the site at an elevation above potential flooding? (Are logs or driftwood on the development site?)
3. Does the site face a relatively sheltered body of water? (Sites facing a large stretch of open water should be at a higher elevation, and buildings should be set well back from the shore.)
4. Do the beach and shore bluff show signs of active and sustained erosion such as tree roots hanging over the bluff edge, soil and rock debris recently fallen onto the beach, or remnants of destroyed seawalls on the beach?
5. Does the beach appear wide and stable with dune grass growing on the upper beach at the base of a bluff?
6. Does the bluff appear stable with little evidence of sliding and a solid cover of vegetation on the bluff face?
7. Are streams or seeping spring water actively flowing down and out from the face of the bluff?
8. Are there signs of active gully erosion down the bluff face?
9. Do neighboring properties have retaining walls or seawalls? (If nearby properties have seawalls or revetment—riprap—in place, inquire about the nature of the problem and about individual experiences.)
10. Are seawalls and groins present? (If they are, expect the worst.

Neighbors have become greedy in their desire to "capture a beach" at the expense of one another.)

11. Does the sediment supply to the beach appear to be uninterrupted by man-made structures such as groins?
12. Have nearby homes on similar sites been there a long time without damage from flooding or wave erosion? (Ask questions to find out.)
13. Are soil and elevation suitable for septic tank operation?
14. Is federal flood insurance available? (Check with local planning officials.)
15. Is there a good supply of drinking water available?

5 The risk classification—quantifying the subjective

We believe that the maps and discussion in this chapter are the most important parts of this book. By disseminating this information we hope to accomplish several things. First of all, the concerned property owner or potential owner will have a basis for a better understanding of regional shoreline hazards. We hope that the book will contribute to a decision to buy or not to buy, to build or not to build. We also hope that this book will spark a better public understanding of the problems and potentials of the Puget Sound and Georgia Strait shorelines. We particularly hope the public will understand that many of their beaches are being degraded and even destroyed in many areas by careless development techniques and the installation of seawalls and groins. Public support of a strong coastal management program is important for the future of Washington's marine shoreline, especially if we want our grandchildren to enjoy it as we have done. The North Carolina book, *From Currituck to Calabash: Living with North Carolina's Barrier Islands*, the oldest in this *Living with the Shore* series, played an important role in molding public opinion and gathering support for recent regulations forbidding all future construction of seawalls along the North Carolina shore. While this kind of legislation might not be appropriate for Washington State, we hope this book will be equally successful in raising the level of public awareness about Washington's inland marine shoreline.

In the maps that follow we have classified all of the Puget Sound and Georgia Strait shoreline into one of three risk categories: high, moderate, and low. It is a subjective classification. No two coastal scientists would come up with exactly the same classifications.

The classifications are general both in detail and scale and should be treated as such. Specific information about a particular region or site would have to come from a much more detailed on-site investigation.

Many times throughout the book we make the point that potential buyers of shorefront property should consider hiring a specialist to analyze the site for potential development problems prior to making a substantial financial outlay. Another possibility is to request a geologic report from the seller as a condition of the sale.

The classification system we used is based on the three principal hazards that are most frequently found along the coastal zone of Washington State. These hazards are beach and shore bluff erosion, landsliding, and flooding. In addition, we consider such things as the likelihood of future beach or shore problems and the probability that a landowner will be forced into the expense of installing some kind of erosion protection structure.

In summary:

High Risk: Hazardous Processes Are Active.
Landslides, shoreline erosion and flooding, or a combination of these hazards are an active threat now and into the foreseeable future.

– High risk of land loss and structural damage.
– If development is present, seawalls, rock riprap, and groins are or will be installed. Such structural defenses against the sea are both expensive and undesirable because of the high cost of construction and maintenance and because the degradation of the beach will continue to the point of eventual loss.

Moderate Risk: Hazardous Processes Are Present at Reduced Levels.
There is evidence of erosion and/or landsliding but with no immediate threat to land or structures.

– Prudent development practices are imperative to prevent any man-made acceleration of the erosion or landsliding.

Low Risk: Hazardous Processes Are Not Present.
There is little or no evidence of erosion, landsliding, or flooding.

– Good development practices that will not initiate any potential hazards are encouraged.

The following classification scheme is dynamic, for it also takes into account how prudently local people are developing their shoreline. Certainly one of the principal aims of this book is to inform the public of the potential risks and to encourage the kinds of development practices that will at least not worsen the risks and might possibly reduce them. Bear in mind, however, that this rating system is only a guide and not intended to frighten those who find their particular property lying within a high-risk location. It is intended to be informative, much like a sign along a highway informing drivers of potential landsliding or rocks on the roadway. Proceed, but do so with caution. We present these classifications urging people to use the coastal zone with common sense and sound development practices.

Whatcom County

The shoreline of Whatcom County is shaped by southerly winter storm waves as well as by northwesterly waves that develop many tens of kilometers up Georgia Strait in Canada. There are few truly sheltered shores in the county; even Drayton Harbor and Birch Bay occasionally experience high waves and flooding hazards to waterfront development.

Point Roberts

The Point Roberts Peninsula (fig. 5.1) is open to relatively high waves from just about all directions, but particularly from the south and southeast. These waves lead to active erosion and localized flooding problems. Particularly vulnerable are the high, unstable shore bluffs on the northwestern and southeastern sides of the peninsula. The remaining area of the shoreline is relatively more stable but also experiences periods of erosion and flooding, particularly when southerly storms coincide with high tides. Such exceptionally high water levels allow waves to rise higher on the beaches and threaten coastal structures. Homes and cottages along the low beaches to the south and west have installed seawalls and various other types of erosion protection structures. The best protection, as always, is a substantial building setback from the beach.

Semiahmoo Spit

Semiahmoo Spit (fig. 5.2) near Blaine, Washington, is a classic coastal spit created by the long-term deposition of eroded sand and rock derived from Birch Point to the south (fig. 5.3). The southern narrow part of the spit is very susceptible to overwash by storm waves. Toward the north, the spit widens, but flooding is still a potential problem over the entire spit area. The spit, longtime home of the Alaska Packers Association, is being converted into a large-scale residential and commercial complex. The shoreline should remain relatively stable as long as there is no

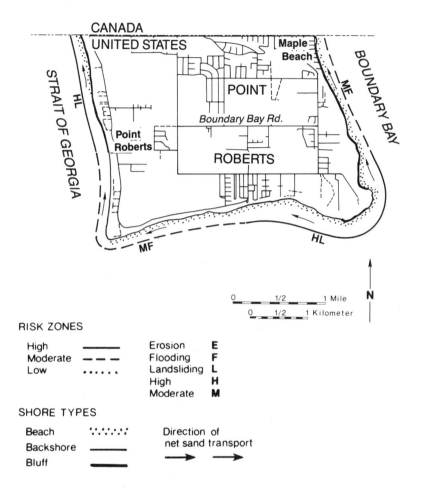

Figure 5.1 Site analysis: Point Roberts.

interruption in the sediment supply to the spit from Birch Point. Spits and other low depositional coastal features are particularly attractive because of their beach access and ease of development. However, it is these very features that make such an area the most vulnerable to erosion and flooding problems. Great care should be taken with their development to reduce potential hazards that can become compounded with time. Exceptionally wide building setbacks are recommended, and seawalls or other erosion protection structures should be prohibited except in cases of "clear and present danger."

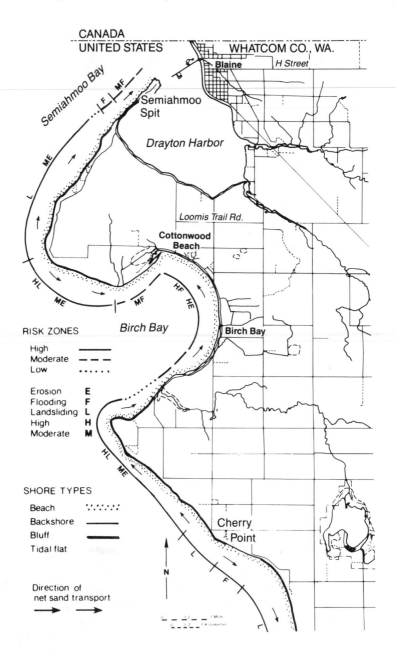

CANADA
UNITED STATES   WHATCOM CO., WA.

Blaine   H Street

Semiahmoo Bay

Semiahmoo
Spit

Drayton Harbor

Loomis Trail Rd.

Cottonwood
Beach

RISK ZONES   *Birch Bay*   Birch Bay

High   ———
Moderate   — — —
Low   · · · · · ·

Erosion   **E**
Flooding   **F**
Landsliding **L**
High   **H**
Moderate   **M**

SHORE TYPES

Beach   · · · · · ·
Backshore   ———
Bluff   ▬▬▬
Tidal flat

Direction of
net sand transport
→   →

Cherry
Point

N

Figure 5.2   Site analysis: Semiahmoo Spit to Cherry Point.

Figure 5.3 Semiahmoo Spit near Blaine is a classic example of a spit that has attracted large-scale development. Proper stewardship will reduce the risk of flooding and erosion.

Birch Point

The shoreline consists of high, eroding shore bluffs with narrow, rocky beaches. It is an important source area not only for sediment that supplies the beach of Semiahmoo Spit to the north, but also for beach sediments that are carried by wave action toward Birch Bay. As such, the Birch Point shoreline is a high-risk development area. Shore bluff erosion and landsliding are common to the entire area. Any development in this region should be required to be built well back from the edge of the shore bluffs to allow natural erosion to continue without threatening coastal structures.

Birch Bay

Birch Bay appears to be a relatively placid, sheltered environment but has a dynamic and interesting shoreline. Large storm waves can be expected from time to time from the south and west. The new housing on the north side of the bay is built upon a series of small, coalesced spits. Sediment for these spits has come from the longtime erosion of Birch Point. This beach sediment supply has been severed by the marina entrance to the southwest. Fortunately this beach is relatively sheltered from wave activity. Installation of seawalls here will only aggravate the erosion problem. It appears that this beach would be a likely candidate for a beach nourishment program in the future.

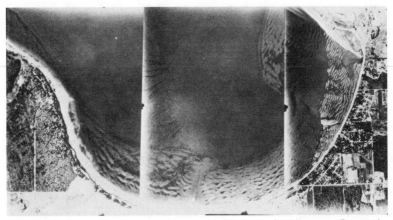

Figure 5.4   Poor shoreline development at Birch Bay in Whatcom County has led to progressive "hardening" of the shore behind seawalls and riprap. Storms coinciding with high tides lead to flooding. North is to the right.

The semicircular sweep of Birch Bay has had a long history of beach erosion and episodic flooding (fig. 5.4). Most of the shore is densely developed, and nearly all of the shorefront homes have seawalls. North of the village, groins have been placed on the beach to help trap the beach sediment and in turn protect the road paralleling the shore. The seawalls and groins have "hardened" the beach, creating what resembles a concrete-defended World War II beachhead. It is clear that none of this would have been needed if, sometime in the past, prudent development practices had been followed. Homes were constructed too close to the water's edge. Unusual storm events, perhaps recurring every five or ten years, caused homeowners to install concrete protection of one type or another that in the long run aggravated the problem.

Point Whitehorn to Neptune Beach

Very high shore bluffs with narrow cobble and gravel beaches dominate this segment of the shore. The bluffs are almost exclusively composed of unconsolidated glacial sediments prone to sliding when saturated. Both landsliding and bluff erosion are found in varying degrees throughout this length of shoreline, making this a relatively high-risk environment for development. Two oil refineries and a large aluminum plant, choosing to locate here because of deepwater access, occupy the uplands. This area will likely see additional marine-related heavy-industrial development in the future.

Sandy Point

Sandy Point (fig. 5.5) is a large spit that has developed from sediments carried by southerly littoral drift currents in the surf zone. The beach is relatively wide and steep but narrows to the south. Homes line the shore, and nearly all these homeowners have installed seawalls of various designs. Most were built following the severe erosion and flooding of April 1977. Additional damage in this storm was caused by drift logs thrown by high storm waves into the waterfront homes.

The seawalls will provide some measure of protection, but beach narrowing is likely to continue, particularly at the southern end of the spit adjacent to the inlet. The future of the beach here is also related to long-planned inlet dredging and stabilization.

Lummi Bay

Lummi Bay is a large tidal flat, formerly the delta for the Nooksack River. The Lummi Indian tribe has constructed a large breakwater here for fish rearing and aquaculture. The west-facing shore of the peninsula to Gooseberry Point has a relatively wide beach, supplied with sediment by actively eroding bluffs. Gooseberry Point is a low depositional site supplied with sediments from those same eroding bluffs. Its low elevation makes it prone to storm-surge flooding.

The beach at Fisherman's Cove is low and shaped by southerly waves that also nourish it with sediment. The shelter of Hale Passage reduces the effects of potential wave erosion, but flooding in the area can be expected during unusual storm events.

Bellingham Bay

The Bellingham Bay shoreline (fig. 5.6) is wide open to the southerly approach of waves that are especially important during the winter. The effects of these waves are clearly seen along the actively eroding shoreline of the Lummi Peninsula. Recession of these bluffs is also affected by water saturation during winter. Riprap has been haphazardly tossed along the shore to help retard the erosion where public utilities and the roadway are threatened. The bluffs are lower in elevation to the east toward the delta of the Nooksack River. This river formerly emptied into Lummi Bay but in the late 1800s shifted its course and has since flowed into Bellingham Bay.

To the east, shore bluffs rise to elevations of 80 feet. Drainage diversions, due to upland development, have caused extensive gully erosion

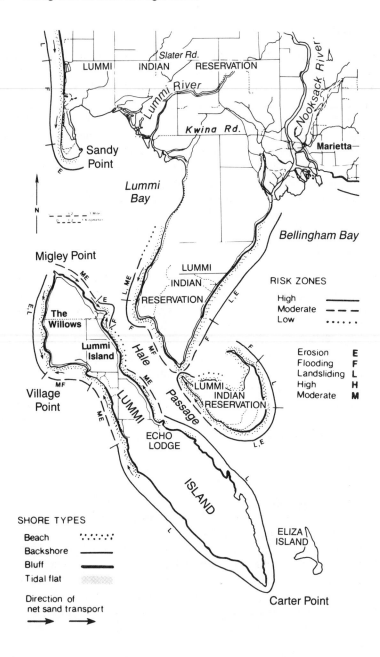

Figure 5.5 Site analysis: Sandy Point to Lummi Island.

and sliding. Prudent upland management of runoff is important to the future stability of these shore bluffs.

Port of Bellingham to Skagit County

The Port of Bellingham is an intensively developed shoreline accommodating industrial, commercial, and recreational activities. Extensive landfills over the decades have completely changed the natural configuration of the shore (fig. 5.7).

One of the few hard-rock shorelines bordering Georgia Strait and Puget Sound is encountered along Bellingham's south side continuing into Skagit County. Steep sandstone cliffs plunge into the water creating a very scenic shoreline. There are many small coves and small pocket beaches in the region, but most of the shore consists of sculpted sandstones pitted by weathering and erosion. The steep topography restricts development to rocky niches and isolated coves. Most development sites are free from wave erosion; however, rock falls are potential hazards.

Lummi Island

Lummi Island is made up of two very distinct environments. The northern half of the island has a relatively low and rolling topography of soft, easily eroded glacial deposits. On the other hand, the southern half is very high and rugged with extremely steep slopes plunging into the water. This high relief reflects a much different, hard-rock foundation. The slopes offer spectacular marine views, but their steepness and ruggedness preclude development.

Point Migley to Lovers' Bluff (west side). Most of the coastal slopes gently rise to shore bluffs up to 20 feet high. The bluffs are eroding and in places are undermining the shore road. Most of the homes are fortunately located behind the road, which creates a natural setback from the eroding bluffs.

The drift of littoral sediments converge at Village Point, creating a low, wide beach curving to Legoe Bay. During big storms, flooding presents a potential hazard to nearby developments. South from Lovers' Bluff the shore is characterized by a narrow gravel beach backed by a 10- to 20-foot high bluff. The entire bluff here is subject to landslide activity, particularly the region between Bakers Reef and Carter Point (south tip of the island).

Point Migley to Lummi Point (east side). The east side of Lummi Island facing Hale Passage is relatively sheltered from storm waves. Narrow gravel beaches with low shore bluffs are common to the area.

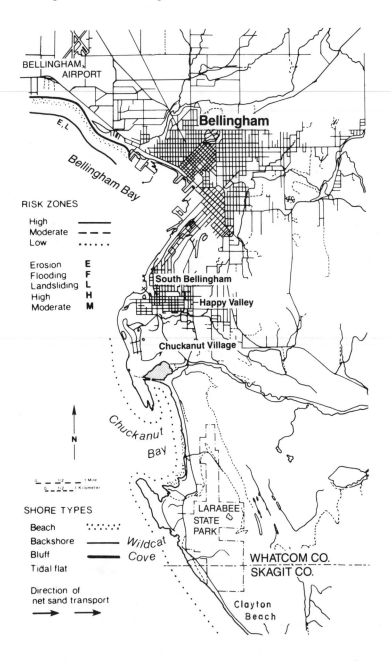

Figure 5.6 Site analysis: Bellingham Bay to Clayton Beach.

Figure 5.7 Unusually high westerly waves strike the riprap shore at Bellingham's Marine Park.

Moderately active bluff erosion occurs all along this side of the shore, but the potential for landsliding is much reduced here compared to the west side of Lummi Island.

Lummi Point is a triangular piece of real estate along this shore where littoral sediment drift converges. This attractive beach has led to relatively dense development along the adjacent shore. Seawalls have been installed here in response to the erosion problem. They will cause further beach degradation and narrowing in the future.

Skagit County

Chuckanut to Samish Island

The steep sandstone cliffs of the Chuckanut area (fig. 5.8) are well known for their sculpted beauty. The sandstone is quite hard and resistant to erosion. However, large slides have occurred in times past along the scenic drive. Residential developments dot the steep slopes, and most are safe from slide hazards. However, some building precautions should be taken to reduce risks. For this reason, it is advisable to have a professional evaluation of a site prior to excavation or filling for site development. Beautiful vistas of the bay and islands make the area extremely appealing, but development is limited by restricted sewer and water supplies.

Samish Island is another popular site for residential development. It was, in the long past, an island, but as the Skagit and Samish River deltas advanced westward, it soon provided a low-elevation, dry-land

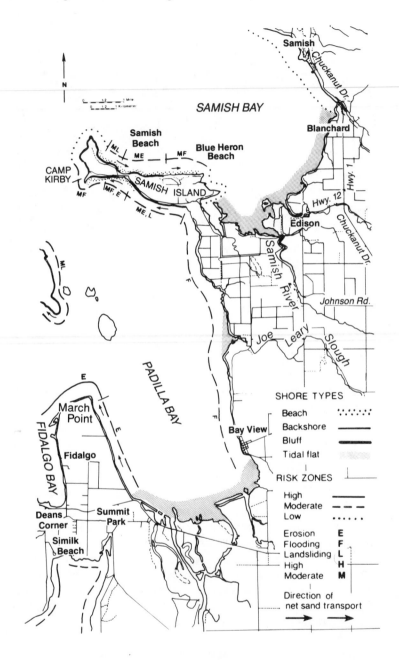

Figure 5.8   Site analysis: Samish Bay to Padilla Bay.

connection to the "mainland." The north side of Samish Island has a number of permanently and seasonally occupied shorefront homes. Most have relatively easy beach access. The south side has much steeper shore bluffs and fewer homes. The shore bluffs on both sides of the island are subject to moderate erosion and sliding.

Any development along Samish Beach should plan for a moderate erosion or landsliding problem. The western tip of the island is composed of bedrock and is essentially free of erosion hazards. Camp Kirby, on the southwest side of the island, is a large body of sand and gravel formed by converging littoral sediments carried from both the north and south.

The long stretch of low-lying, flat shoreline between Blanchard and Samish Island is all Samish River delta front. This area is subjected to shoreline erosion in some areas, but the most severe hazard is flooding by water pushed inland by a storm from the north (storm surge).

Guemes, Sinclair, and Cypress islands

Guemes Island (fig. 5.9) is the most developed of these three neighboring islands. Nearly the entire shore is composed of shore bluffs with variably sized beach aprons leading to the beach. All of the bluffs are subject to moderate or high landsliding risk. The bluffs just north of Kelly's Point, on the southwest side of the island, and at Clark Point, on the north, are the most unstable. The areas with relatively wide beaches and backshores are usually associated with marshes, where there is some possibility of flooding during storms. Several shorefront homes have constructed seawalls or other erosion protection structures. A construction setback, rather than the building of structures, is well advised.

Sinclair Island is less developed. Scattered developments are found along the northern and eastern sides of the island. The eastern shore bluffs are most unstable. The remainder of the shore is relatively stable with pockets of erosion and moderate sliding. There is a potential for storm-surge flooding along the low beach and backshore just west of Urban.

Cypress Island is the least developed and has the most rugged topography of the three islands. Its ruggedness reflects the great extent of bedrock exposed on its surface. Nearly all of the shore is very steep with narrow, rocky beaches. Not only are the bluffs steep, but they are mostly unstable.

In general, the largest storm waves on all three islands can be expected on southwest- and southeast-facing shores.

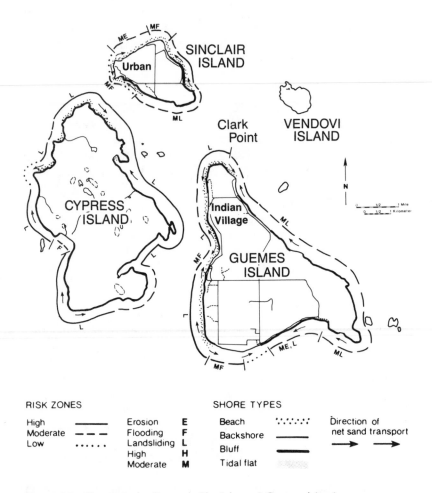

Figure 5.9 Site analysis: Guemes, Sinclair, and Cypress islands.

Padilla Bay and March Point

Padilla Bay is a National Estuarine Sanctuary. This status will protect it from some of the excesses of development in the future. The shore south from Samish Island is composed of "soft" deltaic sands and muds that have been diked to reclaim valuable farmland. Here, the flooding potential during storms is very high. At Bay View the topography rises, and bluffs border the shore. Beyond, the entire shore becomes low and marshy and is subjected to both storm flooding and shoreline erosion.

The eastern shore of March Point borders Padilla Bay. It has low shore bluffs and narrow beaches and is rarely subjected to large waves. In spite of the relatively low wave energies, bluff erosion does occur. Sliding is evident, and most of the northern shore is shrouded behind bulkheads that help support the sliding bluffs. Any shorefront development should be built sufficiently back from the edge of the bluffs to guard the terminus for littoral drift. South from the spit, around the shoreline of Fidalgo Bay, the shoreline is predominately marshy and is subjected to low wave energies.

Fidalgo Island (Anacortes to Deception Pass)

The shoreline around the City of Anacortes (fig. 5.10) is highly altered to accommodate urban and port activities. West from Ship Harbor, the shoreline is composed of narrow rock beaches with low, eroding shore bluffs. The erosion rate here is moderate but should be a factor in any development in the area. These same conditions are largely true for the shore around Washington Park.

Skyline Marina, at the north end of Burrows Bay, is an interesting case where man's development has modified the shore. The net sediment transport around the bay is counterclockwise, and sand flows from east to west in front of the marina. A natural spit grew in the same direction. The developers, for whatever reason, dredged a channel opening through the eastern end of the spit and closed off the natural western opening. The net result is continual channel and spit maintenance problems. This development's approach flies in the face of the natural shoreline dynamics and is clearly not a case of living harmoniously with the shore. Better (and cheaper) to have kept the natural inlet.

The beaches of Burrows Bay are fed by active landsliding down the steep slopes north of Biz Point. Because of recent slide activity, development in this area should be restricted.

South from Biz Point there is a significant change in geology from unconsolidated glacial material to hard, granitelike rock. Even so, some development precautions should be taken on the steep shore cliffs. However, landsliding is much less of a threat. Bowman Bay and Rosario Beach are two small pocket beaches formed in this steep coastal area.

East from Deception Pass the coastal topography remains high with steep shore bluffs. The bluffs are subject to massive sliding, and prior to any development the landslide potential should be investigated (fig. 5.11).

Residential development increases from Yokeko Point eastward to Gibraltar. Narrow, rocky beaches are backed by low, unstable shore bluffs. Much of the instability is due to water saturation of the clay-rich glacial materials in the bluffs. Many of the properties have installed

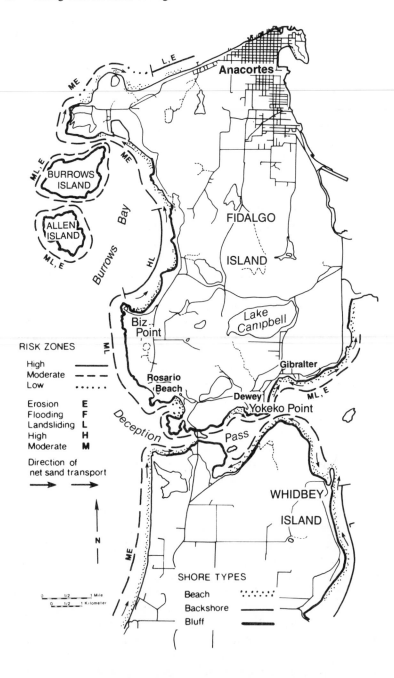

Figure 5.10   Site analysis: Fidalgo Island and Deception Pass.

retaining walls along the shore, clearly indicating the seriousness of this instability problem.

Across Similk Bay, the relatively sheltered shore composed of mud flats and steep, well-vegetated bluffs is also subject to sliding. Similar conditions are found south of Kiket Island (fig. 5.12). Wave energy does increase here, posing some erosion problems. Overall, the sliding and erosion conditions are not severe; nonetheless, all developments should be set back sufficiently from the bluff edge to accommodate the natural recession of the shore.

San Juan County

Because of the great number and the geologic complexity of the San Juan Islands, we would have to write a separate book to cover the coastal hazards in the detail we have covered other areas of Puget Sound and Georgia Strait. Therefore, we will discuss the problems and the joys of San Juan Island living in general terms, leaving it up to individuals to apply the rules of shoreline geology and common sense to their particular situation.

The San Juan Islands *are* San Juan County and vice versa. The principal islands are San Juan, Orcas, and Lopez. There are dozens of smaller islands, including Decatur, Blakely, Shaw, and Waldron islands (fig. 5.13). All of the islands are rugged and beautiful. The past glacial history of the islands can be discerned from their general orientation, and particularly from the orientation of sounds and bays. The shape of

Figure 5.11 Picturesque Deception Pass is one of the few hard-rock shorelines that in spite of its topographic relief is relatively stable.

the islands clearly shows that they were recently grooved and gouged by a giant glacier moving from the northwest to the southeast. The linear nature of the islands becomes even more pronounced as one moves up Georgia Strait into Canada. The Canadian Islands such as Saltspring, Galiano, and Saturna are still largely bare rock. The glaciers left much less material behind in Canada than they did in the San Juan Islands. Unfortunately, since sediment laid down by glaciers is relatively easily

Figure 5.12   Site analysis: Kiket Island to Fir Island.

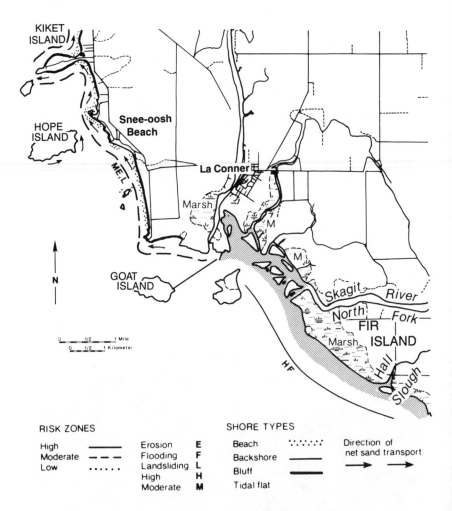

RISK ZONES

High	——————
Moderate	— — —
Low	• • • • • •

Erosion	E
Flooding	F
Landsliding	L
High	H
Moderate	M

SHORE TYPES

Beach	•.•.•.•.•
Backshore	——————
Bluff	██████
Tidal flat	

Direction of
net sand transport
→   →

Figure 5.13 Similar to most of the San Juan Island group, Sucia Island is very rocky. The sculptured shoreline is attractive and relatively safe from erosion and flooding.

eroded by the forces at the shoreline, San Juan islanders must be especially cautious about the erosion problem. Build on rock where possible.

First, we suggest that the interested reader review the general rules for site selection in the previous chapter. These are very much applicable to the San Juan Islands, and all of the Puget Sound shoreline types listed in chapter 4 are found in this county. Perhaps the four most important things a potential San Juan Islands homesite owner might consider are shoreline erosion, island orientation, elevation, and bluff stability.

Shoreline erosion is mainly a problem on the nonrocky shorelines. Beware of undercutting of the upper beach and projecting tree roots from a shore bluff face. Along with the problem of an eroding shoreline comes the problem of what your neighbors are doing about their erosion problem. If they have erected groins or walls in front of eroding bluffs, they have cut off someone's future supply of beach sand and gravel, and that someone could be you.

Orientation of the island shoreline can make a lot of difference in the San Juans. Check to see how exposed your property is to open stretches of water and incoming storm waves. For example, the southern part of Lopez Island and the southwestern tip of San Juan Island face the Strait of Juan de Fuca where large storm waves may be generated. Similarly, places on the north side of Orcas Island may face large storm waves, in this case generated in Georgia Strait. On the other hand, virtually hundreds of coves, large and small, line the shores of these islands, each offering some degree of protection from big storms. East Sound and West Sound, two large indentations in Orcas Island, are the largest bodies of relatively sheltered water in the San Juans.

When thinking about shoreline orientation, assume the worst; that is, assume that your property will one day have to weather a storm striking at highest tide from the direction of the largest stretch of open water. Don't be fooled by the summertime mill pond appearance of many of the harbors in the San Juan Islands. How do you think those big logs got so far up on the shore?

Elevation is something the San Juans have plenty of (fig. 5.14). For example, one would have difficulty finding a point of low elevation on precipitous Blakely Island, although a few homeowners next to Peavine Pass at the north end have managed to build at low elevation.

Most of the other islands vary widely in elevation and relief. Frequently beach-front elevation is quite low at the heads of small bays, and caution should be observed in site selection. Examples of low-elevation bay shores include Shoal Bay and Fisherman Bay on Lopez Island, the head of East Sound on Orcas Island, and Blind Bay on Shaw Island. Because of the elevations on most of the islands, coastal

Figure 5.14 The San Juan Islands are "windows" to the bedrock geology of western Washington.

flooding occurs only in a few isolated locations. Those are generally low, sandy spits of land, which can be attractive building sites. Look for drift logs and reedlike vegetation associated with wet sites. Then check with city hall.

Sites on bluffs of unconsolidated sand and gravel must be viewed with great caution. Set back from the bluff as far as is practicable. Beware of bluffs with exposed tree roots, active springs flowing from the face, unvegetated bluff faces, and bluffs with masses of recently fallen debris (soil, rock, trees) at the base. Any sign of active erosion or slumping on a bluff face is cause for concern and probably avoidance.

Choice of a homesite in the San Juans really boils down to common sense. Stand on the site and go over the checklist in chapter 4. Check city hall for flood maps and advice, and check about storm history with neighbors who live nearby all year round. Buyer beware! Ask the realtor for any hazard information, but don't rely on his or her opinion alone.

The Canadian Islands

Extending northwestward from the San Juan Islands and sandwiched between Georgia Strait and Vancouver Island are the Gulf Islands of Canada. Except for the accident of political boundaries, the Gulf Islands should be simply a part of the San Juans or vice versa, depending on your viewpoint.

The principal Canadian Islands in this group include Saltspring, Galiano, and Gabriola. As mentioned in the San Juan County description, these islands show a very clear northwest-southeast lineation or orientation due to erosion by glaciers. Something like a fossil compass, the grooves gouged out of rock by the glaciers are oriented in the direction in which the ice moved thousands of years ago.

If you're thinking of locating on these Gulf Islands, their glacial history brings you both good and bad news. The bad news is that the glaciers removed much of the topsoil and carried it south into Washington State. The good news is that since the glaciers took away, rather than left behind, their accompanying debris, hard rock and safe building sites abound. Other things being equal and international politics aside, the chances of finding a safe homesite are better here than in the San Juans.

In choosing a site in the Canadian Islands, follow the site selection rules and checklist of chapter 4. As always, apply common sense.

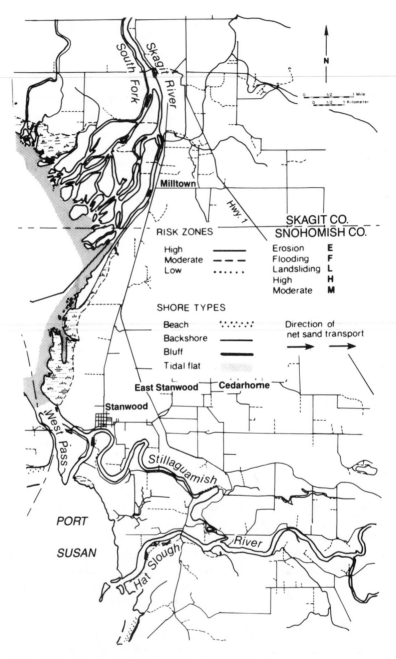

Figure 5.15 Site analysis: Milltown to Port Susan.

Snohomish County

The coast of northern Snohomish County (fig. 5.15) is composed of
low, wide tidal flats formed by the convergence of the Skagit and Still-
aguamish river deltas. Through time, the deltas advanced far enough
westward to create a land connection with Camano Island. Large tracts
of land on the delta have been reclaimed by man for agricultural uses
via levees and tide gates providing some measure of safety from flooding.
The entire shoreline stretch from the county line south to Warm Beach
is highly susceptible to flooding during storms. Just a few years back,
nearly all of the homes at Juniper Beach at the head of Port Susan were
awash due to unusually high water levels.

Warm Beach region

The mainland topography south from Warm Beach (fig. 5.16) rapidly
rises above the floodplain to an undulating, glaciated landscape. The
coast facing Port Susan is dominated by shore bluffs up to 100 feet
high. The beaches are generally narrow except in a few places where
they have widened because of sediment accumulations. The shore bluff
geology changes about halfway between Warm Beach and Kayak Point
from stable bluffs in the north to unstable bluffs in the south. No recent
sliding is obvious, but the potential for sliding is there. Both Kayak
Point and McKees Beach are sites of beach sediment accumulations
that have attracted development. Both sites are vulnerable to flooding
and erosion. Most homeowners have installed seawalls for protection.
The seawalls will provide some protection but are not a long-term solu-
tion. Better to have placed the houses farther back from the shore.

Everett

The entire Everett waterfront from Port Gardner to Elliot Point (fig.
5.17) is modified by heavy industrial development, rail lines, and port
facilities. South from Elliot Point and Mukilteo State Park, the Bur-
lington Northern Rail Line parallels the shore. The line is protected
from erosion by riprap that follows the shore for miles. The uplands
above the tracks reach elevations of greater than 200 feet and are unstable. Between south Everett and Browns Bay, there are many large
sites of active landsliding.

These areas include the communities of Norma Beach and Meadow-
dale (fig. 5.29). Development densities are high here and will likely
increase. Prospective residents or developers should be very aware of

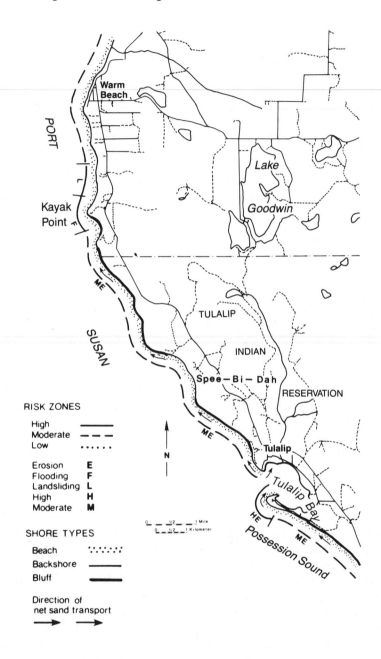

Figure 5.16   Site analysis: Warm Beach to Tulalip Bay.

these unstable slopes and take precautions when buying or building in the area.

South from Browns Bay to the King County line, the shoreline becomes increasingly modified by rail and port/industrial activities. The uplands are relatively stable, save the sloping stretch between Edwards Point and Point Wells.

Island County

Whidbey Island

Island County appropriately consists of two large islands, Camano and Whidbey. Whidbey Island is the largest island in the United States, although New Yorkers from Long Island might dispute that. The beaches along the northwestern part of Whidbey Island are wide and sandy, rather unusual for the beaches in Puget Sound and Georgia Strait (fig. 5.18). The sand for these beaches comes from very high, sandy bluffs several miles to the south. Littoral currents carry the eroded sands to the north toward Deception Pass, along the way supplying the beaches of Deception Pass State Park and West Beach. The littoral currents are driven by relatively large waves that roll out of the Strait of Juan de Fuca, striking the west shore of the island. The high, sandy shore bluffs easily erode, providing constant nourishment for the beaches to the north. These low-lying beaches are exposed to periods of strong westerly winds and waves.

The beach at Swantown, west of Oak Harbor (fig. 5.19), has had a long and sad history of erosion and destruction. The ruined remnants of a huge concrete seawall, broken and fallen on the beach, clearly show the futility of solidifying a mobile shoreline environment (fig. 5.20).

The developer here tried to "create" oceanfront home sites and shelter them behind a massive seawall. The scheme clearly did not work, and several sites were lost. To the north, riprap, bulkheads, and seawalls line the beach, shielding the homes located in the vulnerable area. From the standpoint of the beach user, these walls don't exactly add to the beauty of the shoreline. Furthermore, they are gradually narrowing the beaches.

Immediately south of Swantown, shore bluffs rise abruptly to elevations of over 200 feet. The bluffs consist of alternating layers of sand and gravel laid down long ago by glaciers. This material is a vital source of sediment to the beaches. As such, the bluffs erode quickly and are unwise building sites. Any development proposals along these high bluffs should be reviewed with great caution, not only because of the hazards involved, but also because of the possible effect on the supply of sand to

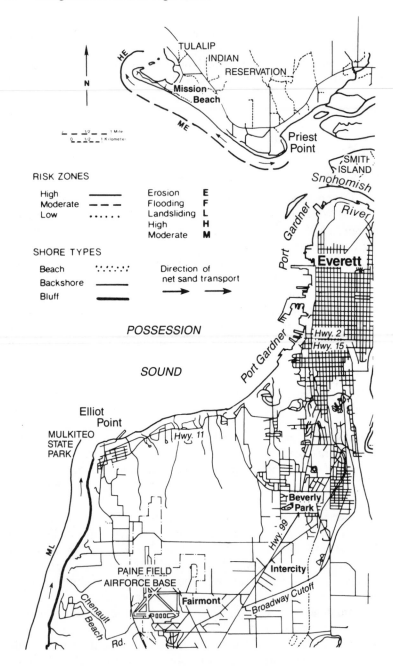

Figure 5.17   Site analysis: Everett and Possession Sound.

Figure 5.18 Facing the Strait of Juan de Fuca, the sand and gravel shore bluffs of Whidbey Island contribute great quantities of sediment to the beaches below.

the beach. The Island County planning and engineering departments are well aware of the hazardous nature of these and other shore bluffs in the county. They have instituted a policy of requiring developers to seek professional advice and prepare a written report evaluating the development risks and mitigating actions.

Partridge Point to Admiralty Head. Similar development caution should be taken along the entire stretch of shore from Partridge Point to Admiralty Head at Fort Casey (fig. 5.21). The bluffs stand at elevations of nearly 200 feet and are experiencing varying degrees of erosional recession. At low tide a wide, healthy beach is exposed at the base of these bluffs. This beach, which is much wider than the beach in front of seawalled communities a few miles to the north, is a product of the eroded bluff materials.

An unusual beach formation called a looped bar is found approximately two miles south of Partridge Point (fig. 5.22). The feature represents the deposition of beach sediment in the form of a bar that loops out from the shore for a distance of several hundred feet. Behind the bar is a low, marshy area subject to flooding.

At Admiralty Head, littoral sediments converge with sediments sweeping westerly around Admiralty Bay but are lost, unfortunately by purely natural processes, to the relatively deep nearshore water. The beach fringing Admiralty Bay is a broad and sweeping one that grew from the progressive transport and deposition of beach sediments from

Figure 5.19   Site analysis: Swantown and Oak Harbor, Whidbey Island.

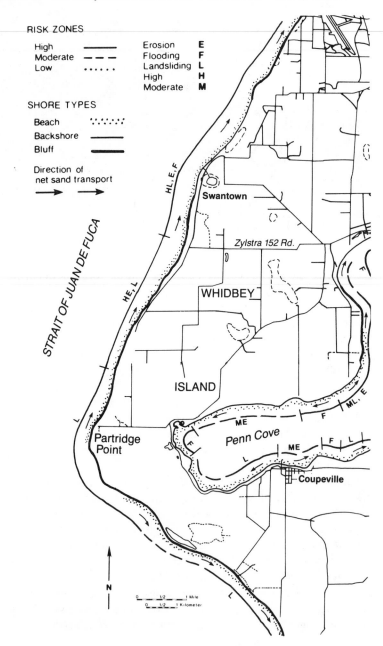

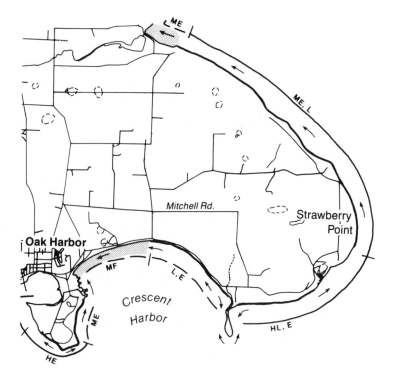

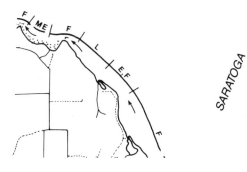

Figure 5.20 Huge, fallen concrete seawalls at Swantown on Whidbey Island point to the problem of trying to solidify the shore and "create" beachfront property.

the south. The shore is directly exposed to southerly storm waves and surges. Some localized flooding does occur; however, the shore is generally high and wide enough to prevent any over-the-road flooding. The entire beach is littered with driftwood, which can serve as a good indicator of maximum high-water flooding limits (fig. 5.23). Any development here should be well back from the upper limit of log debris. This will reduce the likelihood of flooding and flying log damage from high storm waves.

South of Admiralty Bay, coastal topography along the west-facing shore is characterized by 100-foot-high bluffs. These bluffs are the primary source area for the beach sediment along the bay. Suffice it to say that the bluffs are eroding from the base and are prone to landslide failure. Any development along these bluffs should be carefully scru-

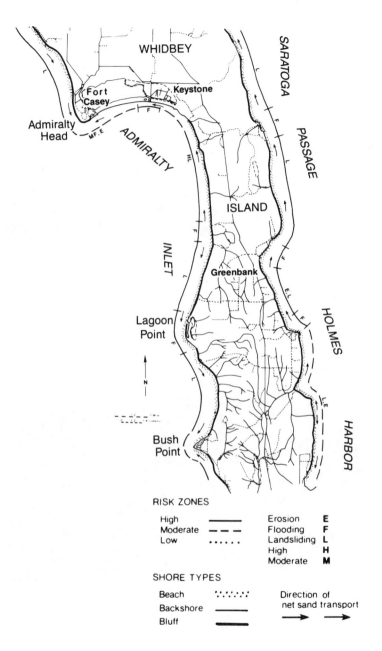

Figure 5.21   Site analysis: Admiralty Head to Bush Point, Whidbey Island.

Figure 5.22 A looped bar extends from the shore at Partridge Point on Whidbey Island. **Figure 5.23** Drift logs are found on most of the region's beaches. Removal is probably unwise as they help trap beach sediment. Yet they can become tools of destruction during storms.

tinized. The evidence of recent and past sliding is abundant. Most of the bluff instability is due to the inherent, weak geological nature of the bluffs; that is, the sand and gravel are unconsolidated. Waves and littoral currents simply sweep away materials that slump down from the bluffs. Landslide-prone conditions prevail all the way down to Mutiny Bay.

The net littoral sediment drift along the west side of Whidbey Island is predominately to the north, but there are local and seasonal variations. The sediments are carried to the north where, in some places, the relatively wide beaches have become attractive development sites. Unfortunately, even these sites are subject to flooding during unusually high-water storm conditions. This is the case at Lagoon Point and Bush Point. Many homeowners have built seawalls for erosion protection.

Mutiny Bay. At Mutiny Bay (fig. 5.24), low, sandy beaches have encouraged a dense housing development. Here, there is no serious landslide problem, but the low coastal area along the shore at Austin is flood-prone. Prior to any substantial investment it would be wise to inquire about past flooding problems and development regulations. Near Double Bluff, the shoreline widens due to an accumulation of beach sediments eroded for a long time from the Double Bluff region. This area has also attracted dense residential development. There is some potential for erosion and flooding, and a wide building setback is recommended.

The high bluffs along both the east and west limbs of Useless Bay are unstable and subject to sliding. Material eroded from these bluffs is carried by littoral drift toward Deer Lagoon. The spit and beach both east and west of the channel are unstable. Residents of Sunlight Beach have installed a series of groins to help trap beach sands. This area is highly vulnerable to both flooding and erosion. Carefully investigate this area prior to any substantial financial investments.

Maxwelton. Maxwelton (fig. 5.25) is located on a low beach platform that has developed from sediments eroded from the high, unstable bluffs toward Scatchet Head. Between Indian Point and Dave Mackie County Park, most of the buildings are in a zone of potential flooding during storms.

East of Scatchet Head the shore bluffs rise to 300 feet and are extremely unstable. Recent landslide activity is clearly evident. An on-site inspection by a professional is recommended prior to investment or site development. Waves carry littoral sediments into Cultus Bay, where a small spit has developed. A dike has been constructed to help reduce flooding. Nonetheless, flooding is still a problem here. On the east side of the bay another spit has developed from the erosion of the bluffs

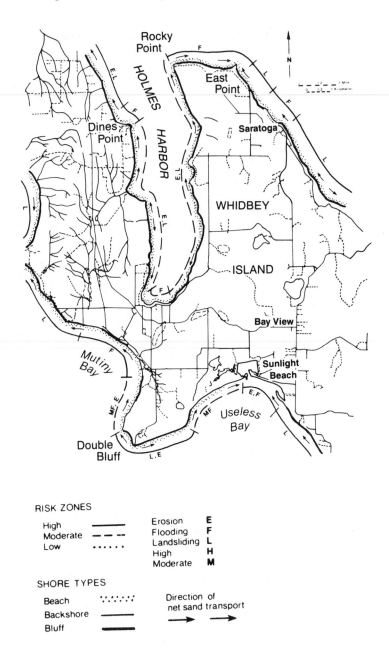

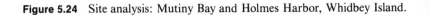

Figure 5.24 Site analysis: Mutiny Bay and Holmes Harbor, Whidbey Island.

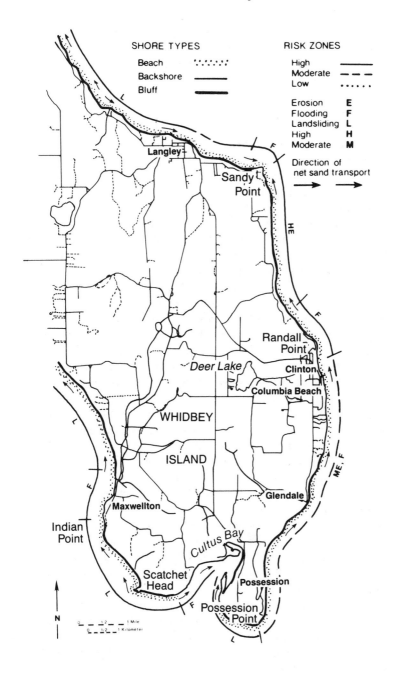

Figure 5.25   Site analysis: Sandy Point to Indian Point.

around Possession Point. It is developed with a small marina. The site does not have a serious flood potential, but shoreline erosion may be a problem, and a careful on-site inspection and inquiry are recommended prior to purchasing property in the area.

Possession and Clinton. Between Possession and Clinton the shore is composed of high, unstable bluffs and narrow beaches that widen in places to accommodate some limited housing. Prevailing southerly waves cause a net northward littoral sediment drift along the shore. In places, the eroding bluffs contribute to the beach sediment, and at Glendale a small stream delivers additional sediment. The low area at Randall Point can experience flooding during unusually high water conditions.

High sedimentary shore bluffs predominate north of Randall Point. They are unstable and show signs of sliding. From Randall Point to Rocky Point, at the entrance to Holmes Harbor, watch out for high, unstable bluffs. The shoreline here is eroding in most, but not all, locations. In addition, there is a potential for flooding along this stretch of the shore where low, wide beaches have developed, such as Randall Point, Sandy Point, Bells Beach, and East Point.

The shoreline around Holmes Harbor is relatively sheltered from storm waves. Nonetheless, there are ample signs of beach and shoreline erosion. Riprap and bulkheads protect several homesites around the bay. The shore bluffs, reaching elevations up to 100 feet, are made of glacial deposits and are generally unstable. A high tide combined with a stiff north wind could cause flooding to low beaches such as Dines Point, Beverly Beach, and the lowlands at the southern end of the bay.

North of Dines Point the high bluffs give way to a low wetland site. The homes here are built behind a large seawall, giving evidence of the flooding problem.

North of Pratts Bluff the shore bluffs are elevated up to 200 feet. Southeasterly waves emanating out of Saratoga Passage strike this shore, adding to the erosional instability of these high bluffs. The spit that has developed across Race Lagoon is a product of eroded sediments from bluff erosion to the south. Flooding occurs along the lower beaches around Race Lagoon. Similarly, Harrington Lagoon has some flooding potential.

Penn Cove, open only to the east, is generally a quiet body of water. Wave erosion is most active at the open end of Penn Cove where wave action is greatest. Shore bluff instability around the cove is spotty, generally coinciding with exposure to storm waves and patches of loosely packed glacial materials.

At the approaches to Oak Harbor the shore bluffs are exposed to storm waves. Those sites most open to southerly storm wave attack suffer the greatest erosion. Maylor Point on the navy base was chosen by the U. S. Army Corps of Engineers to test the durability and efficiency of various long-term erosion protection structures. In 1978 several types of structures using different materials and designs were installed along several hundred feet of Maylor Point (fig. 5.26). The following December a 50-year storm struck western Washington, sinking the Hood Canal Bridge and causing widespread flooding and coastal erosion. The structures at Maylor Point were indeed "tested." The more successful structures were those that had some built-in give, were well drained, and had a filter-cloth backing. The filter cloth helps to retain the sand and gravel backing while allowing water to easily flow out behind the structures. It is certainly worth a trip to Maylor Point to see the "remains" of the erosion structures. The Seattle District of the U. S. Army Corps of Engineers has a lot of detailed information about this project that is available to the public. Use the information that was learned here. (See Appendix of Useful References: Shoreline Erosion and Protection to learn more about this project.)

Beyond Polnell Point, the shoreline of Whidbey Island that faces Skagit Bay rises to high, unstable bluffs. As the shoreline bends toward the northwest beyond Strawberry Island, erosion becomes less of a problem. Unfortunately, bluff instability increases toward Dugualla Bay with many signs of landsliding. Eroded materials from the bluffs are

Figure 5.26 Low-cost erosion control structures were experimentally installed near Oak Harbor by the U.S. Army Corps of Engineers to see which designs and materials were most effective.

carried by littoral action, developing Ben Ure Spit, just west of Hope Island. Toward Deception Pass the shoreline is fairly stable as it is sheltered from storm conditions and in places composed of more resilient geologic materials.

Camano Island

The north end of Camano Island (fig. 5.27) abruptly rises above the Skagit mudflats to elevations in excess of 150 feet. The lowlands west of Stanwood are diked and levied. Nevertheless, flooding prevails during unusual high water conditions. Near Arrowhead Beach, the slopes are stable but do exhibit some signs of sliding and erosion. Arrowhead Beach and Utsaladdy Point are similar as they are both low-accretion beaches, the products of nearby bluff erosion. Both are desirable building sites; however, they are subject to some flooding and beach erosion. Erosion prevails along most of the uplands and shore between the two points. A majority of the property owners have installed erosion protection structures of one sort or another.

Glacial tills intermixed with silts and sands dominate the shore bluff geology around the northwest shoulder of the island and down the entire western shore. Varying degrees of bluff erosion and sliding occur, mainly influenced by subtle changes in the geology and/or orientation of the shore to storm winds and waves. One of the more active landsliding sites is found at Camano Island State Park. Fortunately, it poses no threat to private property.

Clusters of housing tend to be found where the upland topography slopes more gently toward the water, such as at Madrona Beach and Camano and Saratoga Shores (fig. 5.28). Other clusters of homes cling to narrow, isolated beaches such as Indian Beach and Cama Beach. Nearly every one of these settlements have some "hardened" structures for protection against waves and high water. The narrow beaches are particularly vulnerable to flooding, for after all they are products of wave action and littoral processes! Also prone to flooding are lowlands around south-facing bays, such as Elger Bay. Southerly storm waves mounted upon a high tide cause water to pile up, flooding normally dry land. The lowlands at the head of Elger Bay are such a flood-prone site.

The sand and gravel shore bluffs rising south from Elger Bay to Camano Head are uniformly unstable. Varying amounts of erosion and sliding occur on all of the slopes, with the most activity occurring at Camano Head. Here, southerly storm waves focus their attack, directly gnawing at the base of the bluffs or removing debris fallen upon the beach.

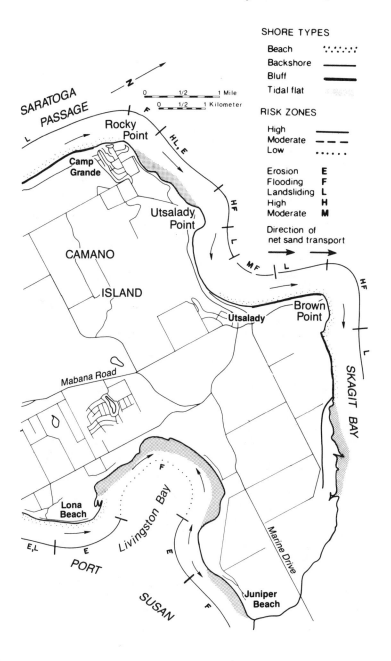

Figure 5.27   Site analysis: Camano Island, northern portion.

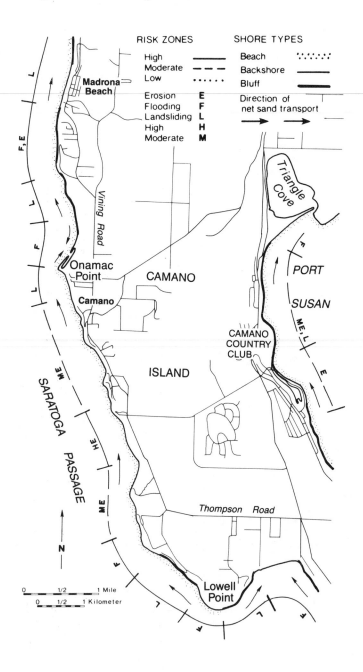

Figure 5.28 Site analysis: Camano Island, southern portion.

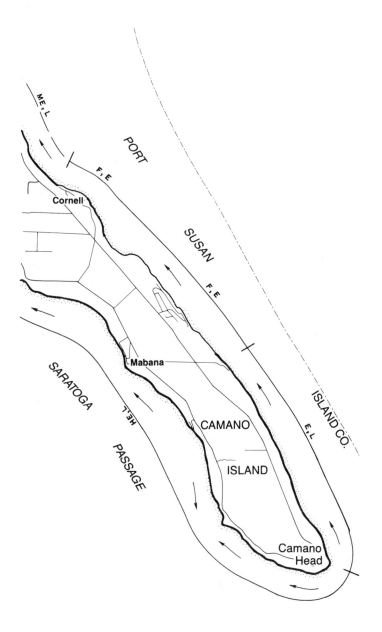

North of Camano Head on the east side of the island, shore bluffs reach elevations in excess of 300 feet. Sands and gravels from these slopes are the source materials for the narrow beaches to the north, Tyee Beach, Tillicum Beach, and Sunny Shore Acres. Seawalls at these beaches provide some measure of protection from wave erosion and flooding, but these are potential problem areas.

Toward the head of Port Susan, lowland sites become even more vulnerable to flood inundation. There is no outlet for storm-driven waters coming out of the south. When storm waters coincide with a lunar high tide, the low places of Driftwood Shores, Lona Beach, and Juniper Beach are very vulnerable. These areas were heavily damaged during the unusually severe winter storms of 1982. Diking has relieved some but not all of the flooding problem.

The shore bluffs at the upper end of Port Susan are erosional. Eroded materials are carried by littoral currents into bays and inlets. Bluff-top developments should take care to build well back from the bluff edge, recognizing that during the economic life of the property, erosion and recession of the bluff will occur. Island County has established building setback guidelines for hazardous shore bluffs. Development proposals are reviewed by the county engineer. Variable building setback limits are related to the level of instability and bluff height.

King County

North King County

A rail line parallels the shore of northern King County and eventually terminates in downtown Seattle (fig. 5.29). Between the north county line and Meadow Point, just north of Shilshole Bay, the tracks are protected by a fitted stone seawall that reaches heights of 4 to 4.5 meters above mean high water. Along this stretch of shoreline, the largest storm waves are likely to approach from the northwest. The seawall has significantly modified the local coastal processes by preventing erosion of the bluffs that, in times past, furnished much beach sediment. The wall acts to protect not only the rail lines, but also helps to stabilize the relatively unstable bluffs found along the northern King County shore. However, the bluffs are particularly unstable south from the highlands to Shilshole Bay and have experienced recent sliding. The sliding problem is inherent in the geology of the area. Layers of clay, gravels, and sand all react differently to ground and surface waters. In January and February when the ground is most saturated with water, slopes are most likely to lose their ability to resist the downward pull of gravity.

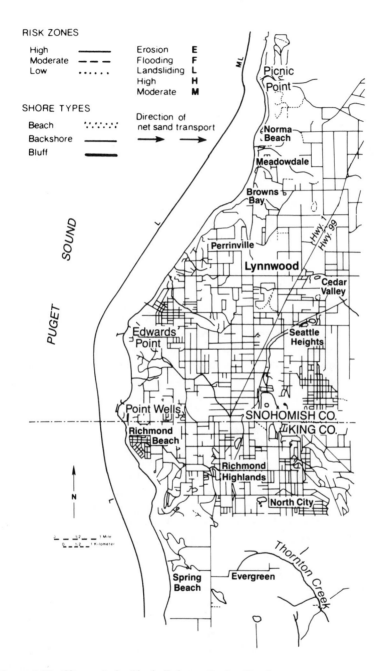

Figure 5.29 Site analysis: Picnic Point to Spring Beach.

South from Meadow Point the shoreline has been filled and land reclaimed from the sea by construction of the Shilshole Marina.

Magnolia Bluff

The Magnolia Bluff (Discovery Park) shoreline (fig. 5.30) is affected by storm waves primarily from the northwest and southwest. The high shore bluffs composed of glacial tills are moderately stable. Many beach-front residents have constructed seawalls to repel the action of the storm waves and halt shoreline retreat. West Point, a low, pointed feature projecting into Puget Sound, is made up of sediment eroded from nearby bluffs and then carried along the beaches.

The combination of relatively unstable bluffs and exposure to southerly storms that cause beach erosion makes the southwest shore of Magnolia Bluff a high-risk zone. Historically, this area has been subject to landsliding. Attempts to halt or slow down erosion have not been altogether successful, as seen by the number of destroyed seawalls and wave defense structures.

The natural shoreline of Elliot Bay has long since disappeared. The shoreline here is entirely modified by man. Some filling to make new land has occurred, particularly at Smith Cove and in the Duwamish area. Despite the fact that the shoreline here is entirely man-made, it is still influenced by the same waves that affect more natural shorelines nearby.

Alki Point and south

The shorelines between Alki Point and Duwamish Head are exposed to northerly waves, but the bigger waves come from the south. Southerly storm waves are refracted or bend around Alki Point, which causes them to lose much of their energy. Beach sand and gravel tend to move in the direction from Alki Point to Duwamish Head. Although the beaches are generally sheltered from big storm waves, beach erosion and some shorefront flooding occur. Nearly the entire shoreline is shrouded with riprap and seawalls, primarily protecting Alki Avenue that parallels the shore. New seawalls are still being built here—for example, in front of the new condominiums near Alki Point (fig. 5.31).

South from Alki Point (fig. 5.32) is a higher-energy shore environment. Southerly winds and storm waves directly strike this northwest-southeast-trending stretch of shore. Continuous bulkheads and seawalls line the shore for miles, giving evidence of a widespread erosion problem. The beaches are often abraded by storm waves that leave behind

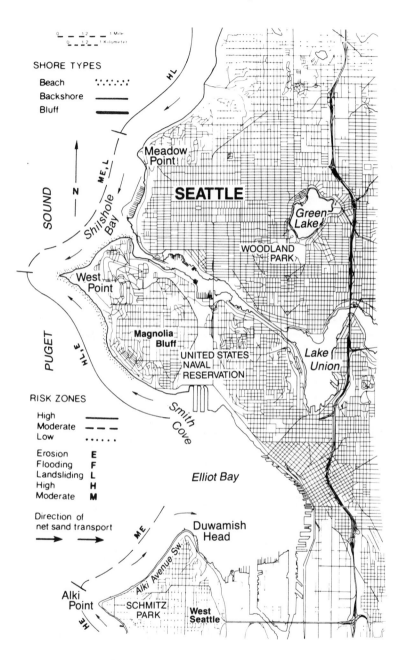

Figure 5.30   Site analysis: Seattle to Alki Point.

Figure 5.31 As the demand for high-density residential waterfront housing continues to grow, greater attention should be given to preserving the shore in a natural state, reducing the need for more seawalls and riprap.

only the largest and heaviest cobbles. Furthermore, the beaches are becoming narrower in front of the seawalls. However, the beaches just north of Brace Point, Williams Point, and Three Tree Point are somewhat sheltered from the storm waves and as a result are wider and composed of more sand.

Many homes are built right next to the shore with bulkheads, seawalls, and/or riprap for protection. It should be remembered that these structures are costly to install and require maintenance, particularly after storms. They are also causing beach degradation, and as the beach narrows in the future, the walls will have to be built even more strongly to withstand storm waves. In areas such as these, the best protection is to set buildings back far from the water's edge.

The shore bluffs south of Alki Point are relatively stable. Most have seawalls or bulkheads near their base, which helps reduce wave erosion. Where there are no structures, the bluffs are eroded and there is some slope failure, such as seen at Ed Munro Park and near Scola Beach. While this might seem to be a problem, the erosion at these sites contributes much needed sediment to the northern beaches. Where walls prevent wave attack on the bluffs, sand starvation of the nearby beaches is beginning.

The bluffs south of Three Tree Point show some signs of instability and old landsliding. Sites oriented so as to be most directly exposed to southerly or southwesterly storms are most vulnerable. The erosion of the bluffs can be viewed as a problem but also as a resource. The eroded

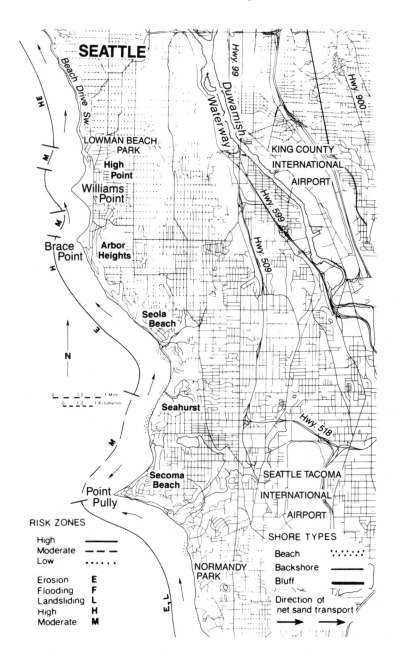

Figure 5.32 Site analysis: Seattle to Point Pully.

sediment falls to the beach, providing sediment. Presumably, as this area becomes developed, owners will install structures along the base of the bluffs to reduce the erosion. Unfortunately this will result in less sediment input, a narrowing of the beach, and great potential for increased erosion damage.

Des Moines

South from the Des Moines Marina and into Poverty Bay (fig. 5.33) the shoreline turns to the southwest. In spite of this shoreline orientation, the largest waves likely to be encountered here will come from storms blowing out of the southwest. The sheltering effect of the shoreline causes a net southerly flow of littoral currents from Saltwater State Park into Poverty Bay and southwesterly to Dash Point. Bulkheads have been installed in the Redondo area where homes and the roads are close to the water's edge. Beyond Redondo, erosion of the unprotected shore bluffs provides sediments to the beaches. Because the shoreline in the area is sheltered from southerly waves and the bluffs freely contribute sediment, the beaches are broad and composed predominately of sand.

Vashon Island

The east- and west-facing shoreline north and south of Vashon is subject to both landslides and erosion (fig. 5.34). Most of the coastal bluffs are unstable and display evidence of recent sliding. Fortunately, along the relatively undeveloped stretches of the shoreline along both sides of the island, the bluffs do erode and feed sediment to the island's beaches. Where beachfront settlements are located, homeowners have built seawalls. In most cases the seawalls and other forms of protection are necessary either because the homes were built close to the water's edge, or the shoreline retreat has "brought" the houses close to the edge (fig. 5.35).

The movement of littoral sediments is generally to the north along the eastern shorelines and to the south along the west side of the island. There are, however, local exceptions to these sand transport directions due to the orientation and exposure of the shoreline to storm waves.

Along the eastern shore of Vashon Island beach sediment is generally moved to the south except in the lee of promontories such as Point Beals and Point Heyes. Robinson Point on Maury Island is a convergence point for beach sediment migrating both from the north and south.

The southeast shore of Maury Island is characterized by shore bluffs

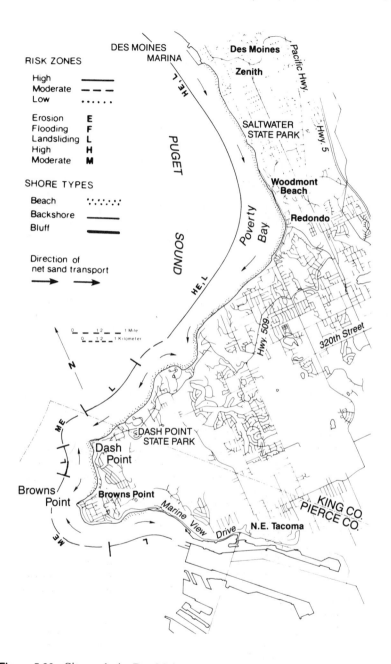

Figure 5.33 Site analysis: Des Moines to northeast Tacoma.

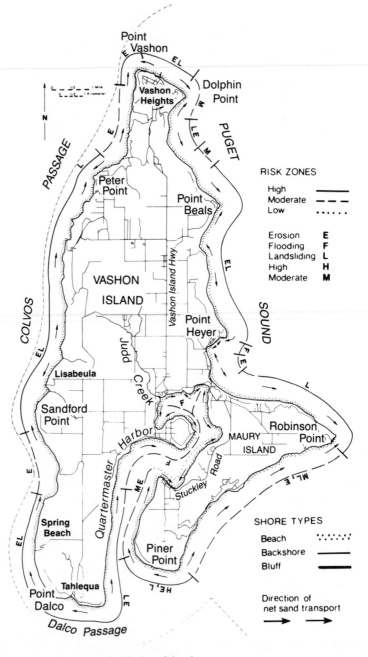

Figure 5.34 Site analysis: Vashon Island.

Figure 5.35   A rare high tide and storm combined to damage several properties at Sandy Shores on Vashon Island (photo courtesy Jay Becker).

up to 300 feet high. In the past, sand and gravel were commercially mined from these bluffs. Localized landsliding does occur, but in general the bluffs are stable. Material that erodes from the bluffs is added to the beach below, providing an adequate supply of beach sediment. The inner waters of Quartermaster Harbor are sheltered from erosion. There is, however, some potential for flooding during storms, particularly along low-lying stretches of shoreline.

The entire south shore and west-facing shore of Vashon from Tahlequah north to Lisabuela is uniformly subjected to landsliding and beach erosion. There are sites where the erosion and landsliding are relatively more active. This occurs because of local changes in the type of bluff material or degree of exposure to high wave conditions.

Beach sediments contributed from shore bluff erosion and small streams are carried to the north by littoral currents. Only at Lisabuela, where there is a slight reorientation of the shoreline, is there a reversal of net northerly littoral flow.

Pierce County

The coastline of Pierce County is typical of most in Puget Sound. It has a meandering configuration with shore bluffs of layered glacial deposits. Beaches tend to be relatively narrow and rocky. The highest-energy waves come from the south, but their full effects are tempered by numerous islands, narrow channels, and a sinuous shoreline.

Commencement Bay—Point Defiance

The promontory, north of Commencement Bay (fig. 5.36), consists of both high, steep shore bluffs and gently sloping, lower-elevation coastal topography. Where the bluffs are steep and high, they are subject to occasional sliding. The densest developments are found where the topography gently lowers to the shore at Dash Point and Browns Point. Beach erosion in this region is significant enough to prompt most home-owners to harden the shore with seawalls, but storm-surge flooding is not a significant hazard.

Nearly all of the shoreline within Commencement Bay is modified by industrial and port-related structures. Natural shore processes are significantly interrupted by these developments. The Tacoma shoreline facing Commencement Bay has small areas of landslide hazards, but coastal flooding during storms should not be an important hazard.

The natural shore is exposed at Point Defiance Park. Here, the high shore bluffs display dramatic examples of active landsliding (fig. 1.4). Great quantities of sand and gravel slide down the bluffs to the shore. The sediment is then driven to Point Defiance by littoral or surf-zone currents where it is lost to the deep water of the Tacoma Narrows.

The high shore bluffs continue well south of the Point Defiance Park boundary. Immediately to the south of the park is Salmon Beach, a small community of homes occupying the precarious area between the water's edge and high (200 to 300 feet), unstable, landslide-prone shore bluffs (fig. 5.37). The homes are built on piles and defended by seawalls.

Between Salmon Beach and Westward Siding the landsliding potential of Tacoma's shore bluffs remains high. Farther south along this west-facing shoreline (fig. 5.38), there are two other areas of significant landslide hazard. These are the shoreline areas around Hoffman Woods and the stretch between Gordon Point and Solo Point. The beach along this stretch is defended against active erosion by riprap that protects the rail line paralleling the entire shore south to the Nisqually Delta. Seaward of the rail line are a few narrow beaches of irregular width, but sediment delivery to these beaches, normally supplied from eroding bluffs, is almost entirely blocked by the rail line. Net transport of beach sand and gravel is from south to north along the entire shore, although there are a few short reversals of this flow where the alignment of the shore changes. A few small shorefront developments are found on the beaches. Because of the relative narrowness of the beaches and lack of significant sediment availability, these developments all have seawalls and groins to reduce storm wave attack and trap the little beach sediment that is available. Most notable of these residential areas are Days

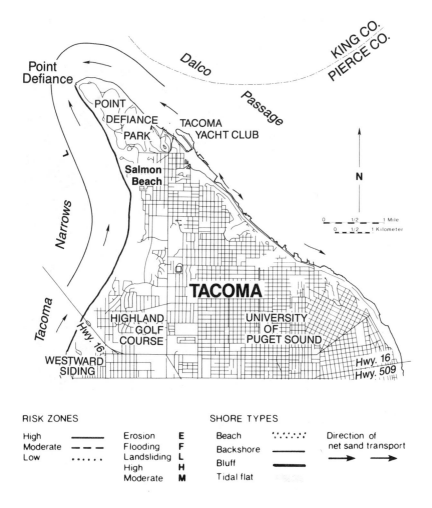

Figure 5.36   Site analysis: Point Defiance and Tacoma.

Figure 5.37 Caught between "water and a high place," homes at Salmon Beach near Tacoma occupy a precarious perch.

Island and Sunset Beach (fig. 3.7). Both communities have homes hidden behind thick and high concrete walls with a few precious pockets of beach sand (fig. 5.39).

The Peninsula

The western side of Pierce County consists of a long peninsula bounded on the east by Carr Inlet and on the west by Case Inlet (fig. 5.40). The shorelines on the peninsula that are exposed to southern storm waves are the most actively eroded. Sediment transport on both the east and west shores moves predominately northward, except where the orientation of the shoreline causes local changes.

High, steeply sloping shore bluffs rim nearly all of the peninsula. The bluffs are made up of glacial deposits, most of which are unstable and subject to slope failure. Care should be taken with building upon high-

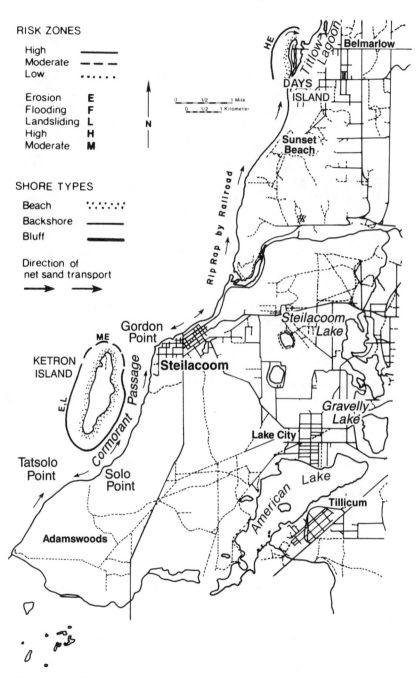

RISK ZONES

High
Moderate
Low

Erosion   E
Flooding   F
Landsliding   L
High   H
Moderate   M

SHORE TYPES

Beach
Backshore
Bluff

Direction of
net sand transport

Figure 5.38   Site analysis: Belmarlow to Tatsolo Point.

Figure 5.39 The moderately high shore bluffs at Titlow Beach near Tacoma are rapidly retreating; fortunately, few homes or structures are in the area.

elevation lots. These bluff-edge sites are most likely candidates for sliding, particularly following long periods of rain in the late winter when the ground is saturated with water. There are many bluff sites that show recent evidence of failure. Failures often take the form of semi-circular notches, easily visible to the eye of any observer who wishes to hike along the bluffs. This kind of slope failure is very difficult to prevent. The best policy is to build back far enough from the edge of a shore bluff so that, should a failure occur, there will be no immediate threat to any structure. Flooding from storm surges is not a problem for the peninsula.

Gig Harbor and vicinity

The shoreline for several miles north and south of Gig Harbor (fig. 5.41) is made up of very high, unstable bluffs. In spite of this fact, the landslide hazard within the harbor is only slight. Erosion and sliding from the Tacoma Narrows north to the Pierce County line are very active, and many houses are already built in active landslide zones. Construction on the narrow beach below the bluffs is not recommended as the beaches are too narrow to accommodate suitable construction sites. Coastal flooding during storms is not likely to be a problem here.

West of Gig Harbor, on the other side of the peninsula, the shoreline

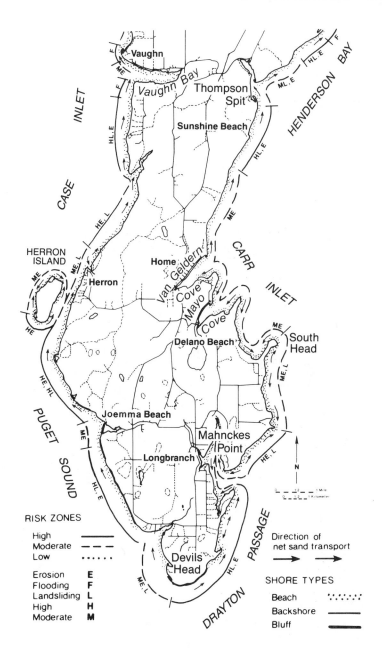

Figure 5.40 Site analysis: Vaughn and Sunshine Beach to Devil's Head.

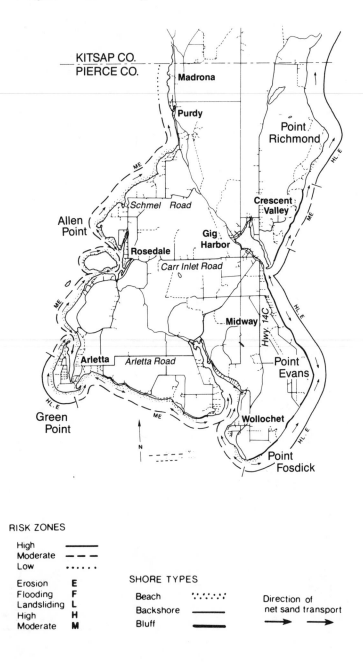

RISK ZONES

High	————
Moderate	— — —
Low	••••••

Erosion	**E**
Flooding	**F**
Landsliding	**L**
High	**H**
Moderate	**M**

SHORE TYPES

Beach	·:·:·:·:
Backshore	————
Bluff	▬▬▬▬

Direction of
net sand transport

→   →

Figure 5.41   Site analysis: Madrona to Point Richmond.

RISK ZONES

High	———
Moderate	— — —
Low	......

Erosion	**E**
Flooding	**F**
Landsliding	**L**
High	**H**
Moderate	**M**

SHORE TYPES

Beach	∵∴∵∴
Backshore	———
Bluff	———
Tidal flat	

Direction of
net sand transport

N

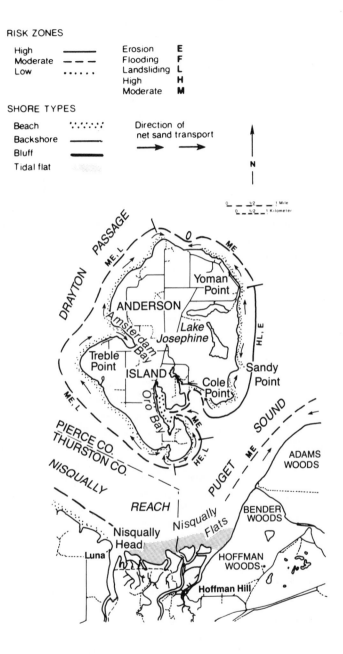

Figure 5.42   Site analysis: Anderson Island.

facing Carr Inlet and Henderson Bay is much more irregular. The coastal topography is rather gentle with only a few steeply sloping shore bluffs. Where there are bluffs, such as the 2.5-mile stretch north of Allen Point, highly unstable landslide conditions prevail. Most coastal development in the area is fronted with seawalls or riprap. The direction of longshore movement of sediment is variable, largely dependent upon the orientation of the shore to prevailing southerly waves.

Anderson Island

Most of the Anderson Island shore (fig. 5.42) is rimmed by shore bluffs 100 to 200 feet high. Almost without exception these bluffs are made up of unconsolidated, unstable material with landslide potential. Only the shorelines of Amsterdam Bay, Oro Bay, and the northeast-facing shore-line along Balch Passage have no significant landslide hazard. Storm waves erode most of the beaches, especially those exposed to direct waves from the south and southwest. Erosion is much less of a problem in the sheltered bays, Oro Bay and East Oro Bay, and the more sheltered shore on the north side of the island.

Fox Island

Exposure to southerly storm waves and high, steep shore bluffs combine to make nearly all of the south and southwest-facing shore highly vulnerable to erosion and landsliding (fig. 5.43). The northwestern side of the island is in the lee of southerly winds and waves; furthermore, it faces the relatively calm water of Hale Passage. This shore, which has few high and unstable bluffs, is most heavily developed. Even on this relatively protected north side of the island, nearly all developments are shrouded behind seawalls, indicating some moderate erosion activity. The future here, like much of developed Puget Sound, promises more seawalls and narrower beaches.

Thurston County

The coastal configuration of Thurston County consists almost exclusively of long peninsulas of land, separated by narrow inlets. This configuration creates many miles of meandering, sheltered coastline. In spite of the relatively protected nature of the shoreline, beach erosion and bluff instability prevail to greater or lesser degrees throughout the coastal region. The Thurston County section of Puget Sound is one of

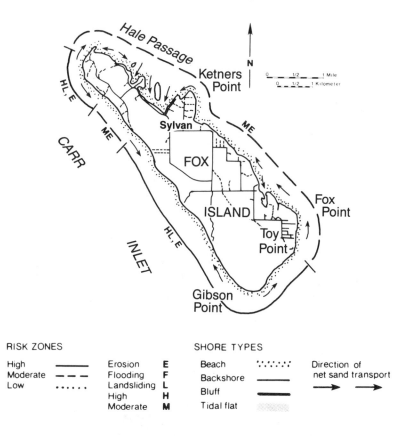

RISK ZONES

High	———	Erosion	**E**
Moderate	— — —	Flooding	**F**
Low	••••••	Landsliding	**L**
		High	**H**
		Moderate	**M**

SHORE TYPES

Beach	·.·.·.·.·
Backshore	———
Bluff	▬▬▬
Tidal flat	░░░

Direction of
net sand transport
→   →

Figure 5.43   Site analysis: Fox Island.

the most highly stabilized by seawalls and other devices. Beach degradation will be the rule of the future (fig. 5.44).

Totten Inlet to Oyster Bay

The shore facing Oyster Bay and Totten Inlet consists of shore bluffs ranging from a few feet to over 100 feet above the water. The bluffs throughout this reach are moderately stable. They are predominately composed of sand and gravel deposits capped with glacial tills. Homesites on the bluffs should be chosen with care, recognizing that erosion and sliding can be a potential problem. Net beach sediment transport driven by southerly waves is almost exclusively toward the northwest. Probably well over 50 percent of the shoreline here is stabilized by walls and the like. Much of the supply of sand from eroding bluffs has

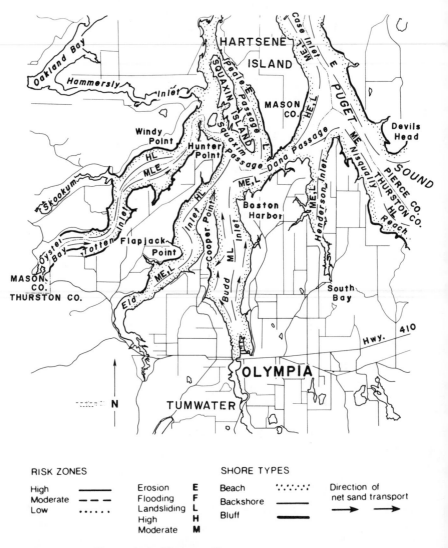

Figure 5.44   Site analysis: Thurston County.

been cut off. The future here will be one of narrowing and disappearing beaches.

Squaxin Passage to Eld Inlet

Southeasterly from Steamboat Island, the land gently slopes to the water with no significant shore bluffs. The beaches receive deflected southerly storm waves and are protected to the north by Hope and Squaxin islands.

South from Hunter Point into Eld Inlet the shore bluffs rise rapidly to elevations up to 100 feet. The bluffs are unstable, and beach erosion is active the entire distance to Frye Cove, just north of Flapjack Point.

Beyond Flapjack Point the south-facing shore is exposed to southerly storm waves. Nearly all of the shorefront homes have installed seawalls. Farther along, the entire west shore of Eld Inlet is also shrouded behind seawalls.

The eastern shore of Eld Inlet terminates at Cooper Point, a distinctive, dagger-shaped spit. The land slopes gently to the shore. Some minor bluff sliding and erosion occur, but these do not appear to be serious problems. Moderate rates of erosion occur along the entire length of the shore to Cooper Point. Seawalls are intermittently installed along the shore.

Budd Inlet

Both the east and west shores of Budd Inlet display moderate amounts of erosion. This is reflected by the number of seawalls built by homeowners (fig. 5.45). There appear to be a greater number of seawalls on the western shore than on the eastern shore; however, this might just indicate a higher density of coastal development on the western shore. Wherever there are bluffs along the inlet shoreline it is safe to assume that there is a significant landslide hazard. On both east and west shores the prevailing beach sediment transport is northerly, save at a few small coves. There is not much likelihood of flood damage to shoreline structures in Budd Inlet, except for some small, low areas in Olympia.

Dana Passage

The shore northeast from Boston Harbor is very irregular with steep shore bluffs in the center, lowering to the east and west. The net sediment movements are directed by the meandering coastline as it inter-

Figure 5.45 Seawalls come in many shapes and sizes. Near Olympia a property owner installed a "designer" type.

cepts the prevailing westerly waves. The high headland promontories tend to have unstable slopes and are the most actively eroded areas. The incidence of erosion quickly changes in the more sheltered locations along the shore toward Henderson Inlet. Much of the shoreline here has been stabilized by walls.

Henderson Inlet

Henderson Inlet is a long, narrow water body flanked by shore bluffs of moderate to low landsliding potential. The narrowness of the inlet restricts the development of high storm waves; nonetheless, moderate erosion does occur along the entire shore. Most shorefront developments have built seawalls to protect their properties from erosion and sliding. The eastern shore is most densely developed and also the most hardened with seawalls. Flood potential during storms is low except for some small areas at the extreme south end of the inlet.

Nisqually Reach

The shore from Johnson Point southeasterly to the Nisqually Delta is backed by shore bluffs up to 100 feet high of moderate stability. Only in one or two isolated locations where the bluffs are very steep is there a high landsliding potential. Beach and shore bluff erosion prevails along the entire reach of the shore, more significant in some places than others. Most beachfront homes have seawalls to retard the loss of their property. With closer proximity to Nisqually Beach, the beach sediment becomes finer and the beach widens, providing greater protection to shorefront homes. Flooding potential is low here except for the low-lying land areas adjacent to the Nisqually Flats (the Nisqually Delta).

Mason County

Hood Canal: Triton Head to Lynch Cove

Hood Canal loops through central Mason County (fig. 5.46). The canal is bordered by slopes that plunge steeply into the water. Narrow gravel and cobble beaches parallel the shore. The western shore of Hood Canal from the Kitsap County line south to the Great Bend is bordered by steep bluffs composed of unconsolidated glacial materials except north of Cummings Point, where ancient volcanic rocks are found. These bluffs exhibit past and present landsliding along essentially the entire western shore.

The eastern shore of Hood Canal is also very steep. The landsliding potential is considerably less, however, as the geologic makeup of the bluffs tends to be more stable. Littoral sediments are driven northward by the prevailing southerly winds and waves. The steep coastal topography prevents most flooding problems; however, some flooding does occur at the deltas of the Hamma Hamma and Skokomish rivers.

The shoreline of Hood Canal, northwesterly from the Great Bend to the end of the canal, is heavily developed, and seawalls and riprap abound. Landslide potential is highly variable here. On the south shore the slopes are largely stable, except east of Forest Beach where instability becomes the rule. The north shore has several broad zones where landslides are a hazard, particularly for a four-mile stretch east of Tahuya. Prudence and professional help are suggested in homesite selection.

Case Inlet

The shoreline of Case Inlet and North Bay is not too heavily developed by southern Puget Sound standards. The beaches are composed of

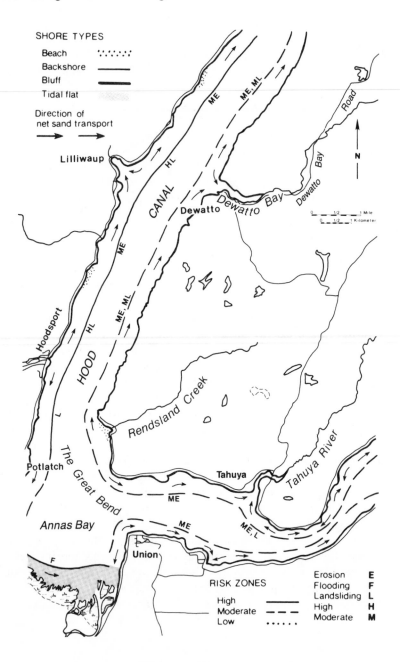

Figure 5.46   Site analysis: Hood Canal, Mason County.

gravels grading to fine sands and mud at the head of North Bay. Moderate shore erosion predominates throughout this area. Erosion protection structures have been installed where developments are clustered along the shore. Similar levels of beach and shore bluff erosion are evident along the shores of nearby Stretch and Reach islands.

Hartstene Island

The entire shoreline around Hartstene Island shows signs of relatively active erosion and landsliding, locally severe in places. The most severe erosion and bluff instability are on the eastern side of the island where larger storm waves are generated along Case Inlet. The narrowness of Peale and Pickering passages on the west side of the island restricts the formation of large storm waves. In spite of the relatively protected environment, erosion and bluff sliding do occur. Moderate beach erosion and bluff instability prevail along the entire western shore of Hartstene Island and just across the passage on the mainland shore. Shorefront developers would be well advised to carefully inspect individual building sites and build structures well back from the shoreline and shore bluffs.

Near Shelton, both Hammersly Inlet and Totten Inlet terminate at long, narrow tidal flats. The shoreline and shore bluffs paralleling these inlets are surprisingly active and show clear signs of erosion and deposition. Tidal currents are particularly strong through these narrow passages and account for some of the processes operating along the shore. Landsliding is particularly active along the shore bluffs southwest from Windy Point along Totten Inlet.

Kitsap County

Western Kitsap County: Hood Canal shoreline

The western side of Kitsap County borders Hood Canal, a very long and narrow water body. Wave energies are relatively restricted because of the narrowness of the canal; nonetheless, shoreline wave erosion and shore bluff landsliding occur on almost the entire length of this shore. Generally neither would have been a significant problem except for the fact that many homeowners who built imprudently close to the water have been forced to build seawalls and place riprap in an attempt to stall erosion. Most of the very steep shore bluffs are subject to sliding, and development should be restricted or carefully monitored. Stretches of shoreline with particularly problematic potential landslide problems include (1) a two-mile stretch south of the Floating Bridge, (2) between

Vinland and Lofall (fig. 5.47), (3) a one-mile stretch north of Warren-ville, (4) the vicinity of Stavis Bay (fig. 5.48), and (5) a two-mile stretch south of Frenchmans Cove (fig. 5.49).

Storm-driven waves that come out of the south drive a predominant northerly flow of littoral sediments. While a south-to-north flow of sedi-ment prevails, it is often interrupted and reversed along several short segments of the shore due to shoreline irregularities. Also, flooding can be a problem at small, low-lying beaches and at the heads of small embayments, particularly where streams enter the shore.

Eroded trees fallen across the beach and numerous groins trapping the flow of sediment point to some of the natural and man-made shore-line problems along Hood Canal. Coastal development and nature can happily coexist here, but one must simply take a few precautions. Care should be taken not to build on low-lying beaches that might flood. Set buildings back from the shoreline with the recognition that natural beach erosion will occur. You don't want to be put into the desperate position of having to try and stop erosion for fear of losing your house or other structures. If possible, don't build on steep slopes. Most of the shore slopes along Hood Canal are moderately stable, but even they can fail under the wrong conditions. Furthermore, building on slopes can alter the natural slopes, soil, and drainage conditions enough to increase the chance of failure.

Sinclair Inlet

The shoreline of Sinclair Inlet (fig. 5.50) is almost entirely modified by industrial development associated with the Puget Sound Naval Shipyard and the riprap-protected rail line to the west. Similarly, the entire shore-line for several miles to the east and west of Port Orchard is modified by riprap. Thus, the natural shoreline is nonexistent, except at the west end of the bay. There is a significant landslide hazard on the steep bluffs facing the water on both sides of this bay. There is no strong likelihood of flooding here.

Dynes Inlet

Dynes Inlet is a relatively small embayment with several small, sheltered bays. The meandering shoreline creates a variety of net sediment trans-port flows as well as erosion and accretion sites. The shoreline is densely developed, and owners in those sections exposed to southerly and north-erly storm waves have been forced to line the shore with seawalls or

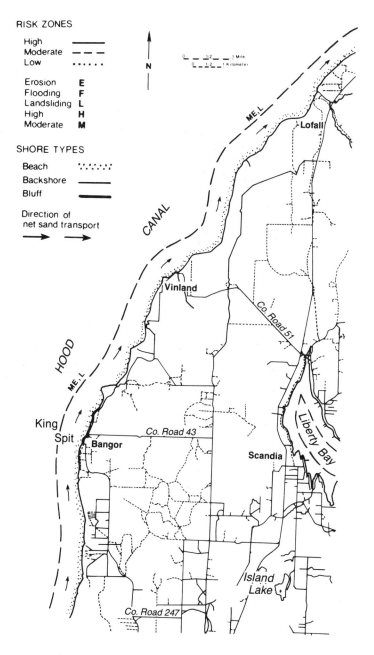

Figure 5.47 Site analysis: Lofall to Bangor.

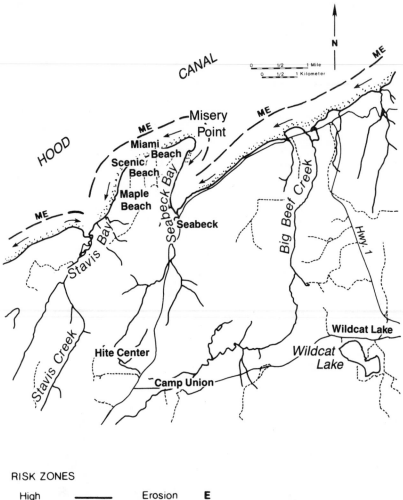

Figure 5.48 Site analysis: Misery Point.

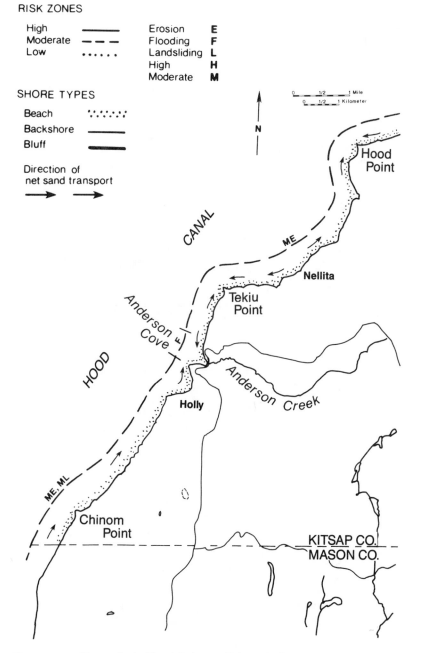

RISK ZONES

High	————	Erosion	**E**
Moderate	– – –	Flooding	**F**
Low	• • • • •	Landsliding	**L**
		High	**H**
		Moderate	**M**

SHORE TYPES

Beach   ∴∴∴∴

Backshore ————

Bluff ━━━━

Direction of
net sand transport

⟶   ⟶

Figure 5.49 Site analysis: Hood Point to Chinom Point.

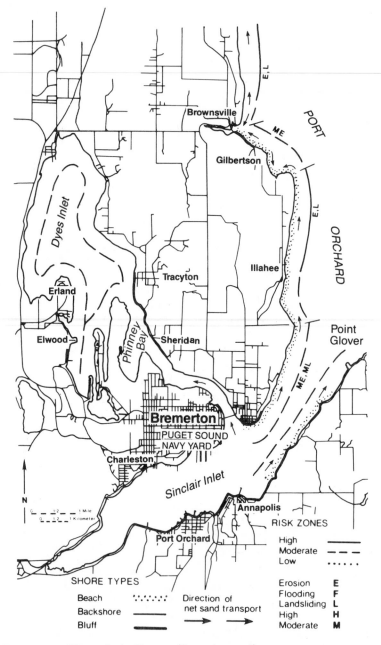

Figure 5.50   Site analysis: Brownsville to Annapolis.

riprap for protection against high waves. The uplands and shore bluffs, composed primarily of glacial tills, are generally stable, although small areas north of Windy Point and at Sulphur Springs have unstable beach-front bluffs. It is difficult to generalize about the shoreline of Dynes Inlet because it is so variable. However, it is classified as a shoreline with moderate erosion potential. Conditions can change from site to site; therefore, a critical inspection of each individual development site is advisable. There is a strong likelihood that beach quality (width, volume) will gradually deteriorate in the future along walled shores.

Port Orchard Channel, west side

Port Orchard is a long, narrow channel separating Bainbridge Island and the Kitsap Peninsula. Its shoreline conditions are shaped by domi-nant southerly winds and waves. The shoreline and shore bluffs on both sides of Port Orchard are erosional. Nearly 50 percent of the shore is hardened with riprap or seawalls. Groins trap some of the north-flowing beach sediment, and this leads to downdrift pockets of erosion. Essen-tially everywhere that there are steep bluffs adjacent to the shore here, the landslide potential is significant.

Liberty Bay

There is a great deal of variability in the direction of sediment flow and shoreline erosion and accretion within Liberty Bay. The sheltered nature of the bay reduces the potential for wave erosion, but nonetheless there are many seawalls lining the shore here. Generally, these small embay-ments tend to trap sediment eroding from nearby beaches, leading to wide beaches and tidal flats. Landslide hazard here is slight except for a few small areas, the most notable being in the vicinity of North Kitsap High School.

Bainbridge Island

The eastern shore of Bainbridge Island (fig. 5.51) is open to southerly and northerly winds and waves from Puget Sound. The southern and western shorelines face smaller water bodies, and the potential for large storm waves is somewhat diminished. The prevailing littoral sediment flow is from south to north, as directed by prevailing southerly waves. Point Monroe is the only heavily developed area with a serious flood hazard. The erosion problem along the island's shore is not severe, but

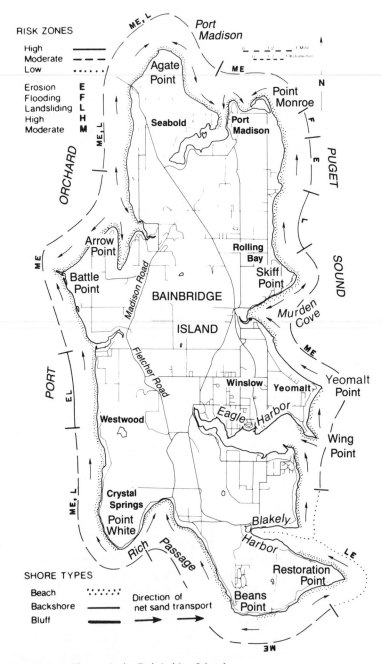

Figure 5.51   Site analysis: Bainbridge Island.

so much construction has been carried out very close to the beach that most beachfront property owners have constructed walls or put in stone riprap. Perhaps just under 50 percent of the Bainbridge Island shoreline is stabilized. Here, as elsewhere in Puget Sound, the high shorefront bluffs are prone to landsliding. Particularly serious landslide hazards are present south of Westwood, at Yemalt, west of Seabold, southeast of Agate Point, and north of Brownsville.

It doesn't take a crystal ball to see that the future of Bainbridge Island holds many more seawalls and riprap structures along the shoreline and that this will be accompanied by a great deal of beach degradation.

Colvos Passage

The Kitsap County shoreline facing Colvos Passage (fig. 5.52) is nearly a mirror image of the western shore of Vashon Island. Both shores have high shore bluffs that are unstable. There are many pockets of recent slide activity. Where beachfront development has occurred, homeowners have built seawalls or placed riprap for protection from storm waves. Shoreline erosion is widespread here, which should encourage building well back from the beach. Development care must be taken in this area, particularly because of the shore bluff sliding problem.

Point No Point to Kingston

Steep coastal bluffs dominate this part of the Kitsap Peninsula (figs. 5.53, 5.54). The bluffs are composed of glacial materials that become unstable, particularly when saturated. There are several locations of recent sliding in the area. Numerous fallen trees on the beach give further evidence of slope instability and beach erosion along this stretch of shoreline. A wide building setback and precautions not to aggravate the landsliding potential of the shore bluffs are sensible approaches to development in the area. Beware of floods during storms at Apple Cove and at Point No Point.

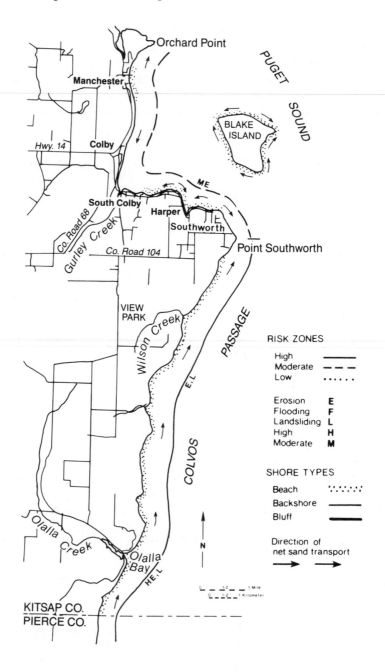

Figure 5.52 Site analysis: Orchard Point to Olalla Bay.

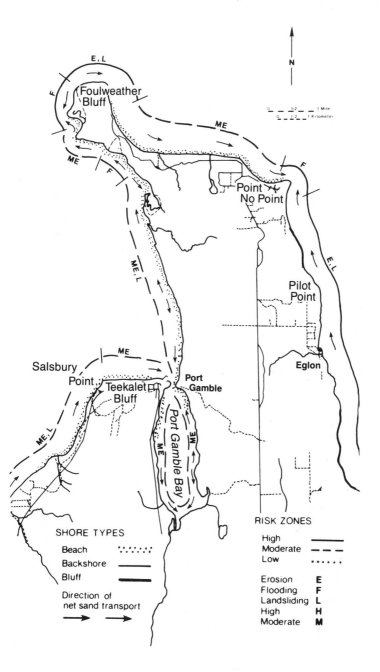

Figure 5.53   Site analysis: Point No Point to Port Gamble Bay.

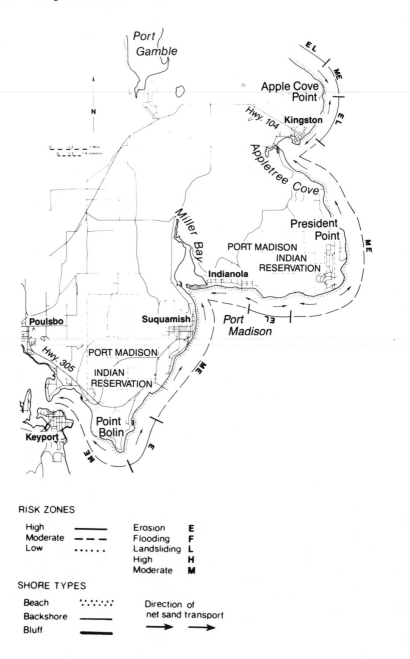

Figure 5.54   Site analysis: Kingston to Poulsbo.

Jefferson County

Steep slopes best characterize the Jefferson County coast bordering on Hood Canal and Admiralty Inlet (fig. 5.55). Some of the slopes are made up of relatively hard volcanic materials; however, most are glacial deposits prone to sliding. It is good practice to assume any coastal bluff in this area has some potential for failure. The degree of risk depends on the steepness and geological composition of the slope as well as its exposure to prevailing storms. There are a few locations of critical sliding (figs. 5.56, 5.57, 5.58, 5.59), the very steep slopes just north of Triton Cove, Thorndyke Bay, and Termination Point north of Squamish Harbor. In addition, sliding is very active around the Toandos Peninsula and many places around Dabob Bay. The shore bluffs west of Fort Worden toward Port Discovery show several signs of failure.

Beach and shore bluff erosion is quite common throughout the region. It tends to be most active along south-facing shores open to high storm waves. The greatest amount of erosion usually occurs when storm waves

Figure 5.55 Steep slopes and narrow beaches best characterize the Hood Canal shore. Isolated accumulations of sediment attract development.

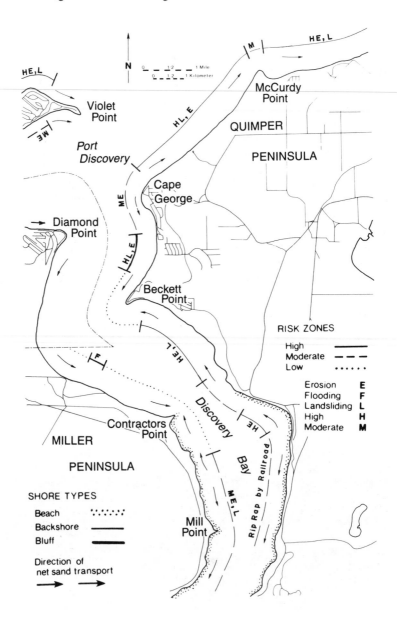

Figure 5.56   Site analysis: Discovery Bay.

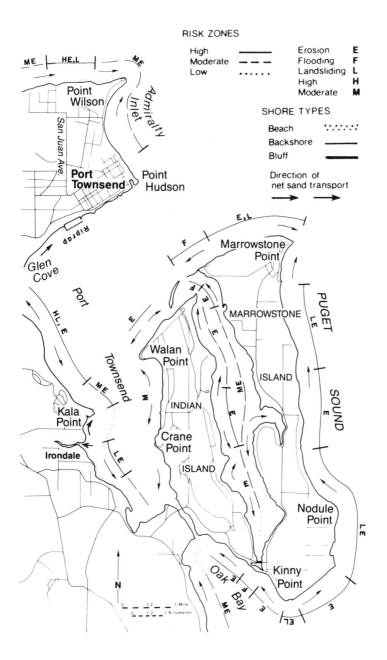

Figure 5.57 Site analysis: Point Wilson to Oak Bay.

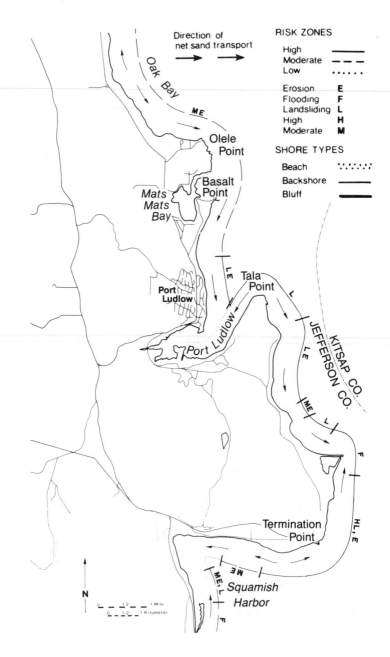

Figure 5.58 Site analysis: Oak Bay to Squamish Harbor.

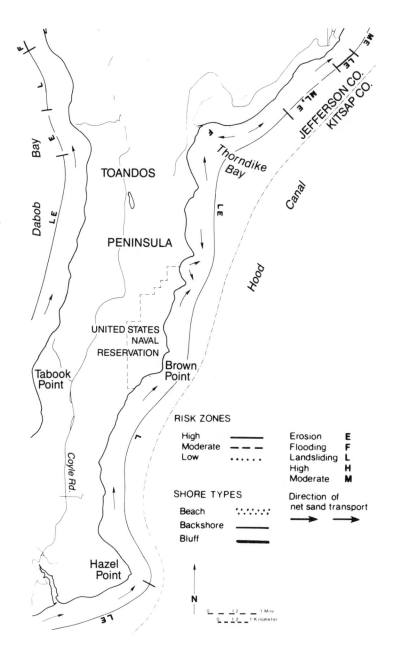

Figure 5.59   Site analysis: Toandos Peninsula.

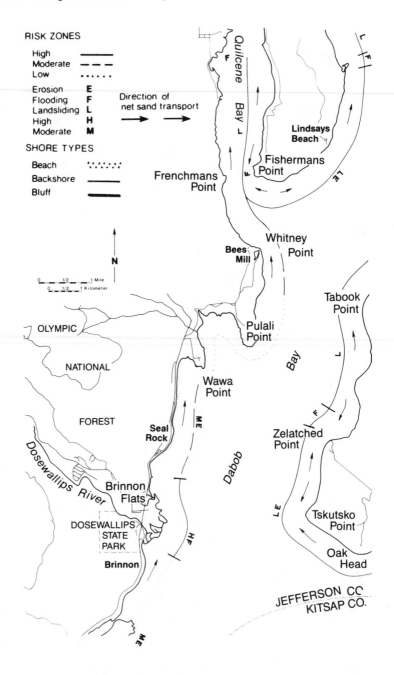

Figure 5.60   Site analysis: Dabob Bay.

Figure 5.61   Fast-moving streams flow down the steep slopes into Hood Canal, and small deltas fan into marine waters. These sites are particularly vulnerable to flooding.

coincide with high tides. Under these conditions, the waves flood the narrow beaches directly attacking the shore bluffs. The most active beach and shore bluff erosion is found at Quatsap Point, Oak Head, Quilcene and Dabob bays (fig. 5.60), Hood Head, and all around Marrowston and Indian Islands. High waves coming out of the Straits of Juan de Fuca erode nearly the entire stretch of shore bluffs west of Fort Worden.

Curiously, in this region of high coastal topography, flooding also presents something of a hazard to development at selected sites. The steep, fast-moving streams emptying into Hood Canal deposit large quantities of deltaic deposits (fig. 5.61). These low sites along Hood Canal are attractive building locations. They are, however, subject to periodic flooding. Some of the higher-risk sites include Duckabush, Brinnon Flats, upper Quilcene and Tarboo bays, and a few lowland sites around Port Discovery and Oak Bay.

6 Coastal land-use planning and regulation

by Peter H. F. Graber

Shoreline management

During the past 15 years, land-use planning and regulation in the Puget Sound area's coastal zone has increased dramatically. The state Shoreline Management Act of 1971 (SMA), which anticipated the federal Coastal Zone Management Act (CZMA) passed the following year, established a framework for the creation of locally initiated master programs that must be consistent with statewide guidelines. Under these programs, a 200-foot-wide strip around the sound is zoned, and permits must be obtained for many types of "substantial development."

The SMA is the cornerstone of the Washington State Coastal Zone Management Program, the first such state program to be approved by the federal government under the CZMA. The National Flood Insurance Program and the U. S. Army Corps of Engineers' dredging and filling permit system are among important federal programs having an impact on coastal land use.

A discussion of some of the relevant federal, state, and local land-use programs, laws, and regulations applicable to the Puget Sound area's coastal zone follows. The explanations provided are introductory and general in nature.

The harbor line system

The Washington Constitution of 1889, still in effect today, created a harbor line system to protect the public interest in Puget Sound and other navigable waters in the state. Article 15 of the Washington Constitution required the legislature to appoint a commission to identify harbor areas and to establish harbor lines within and near those areas.

Legislation in 1927 directed the commission to set both inner and outer harbor lines. Article 15 was amended in 1932 to extend the permissible width of harbors to 2,000 feet. The amendment also permitted the legislature to provide for relocation of harbor lines.

Under Article 15, Washington's harbor areas are "forever reserved for landings, wharves, streets, and other conveniences of navigation and commerce." The term "conveniences of navigation and commerce" is not defined in the constitution, and its meaning has remained in question. The state, as owner of the beds of saltwater harbor areas, is authorized by statute (Revised Code of Washington [RCW] 79.92.460) to lease harbor area lands. By prescribing lease terms and conditions, the state Department of Natural Resources (DNR) exercises control over the private lessees. In addition, the DNR or local port district is empowered to approve plans for any improvements in the leased areas.

Article 15 of the constitution prohibits the state from giving, selling, or leasing to any private person, corporation, or association any rights in the waters beyond the outer harbor lines established under this program.

The state Board of Natural Resources currently administers the harbor line system. It remains the primary mechanism of state control over lands waterward of the line of mean lower low tide in spite of the fact that the state's Shoreline Management Act is the primary method of regulating the area around that line.

Under Article 17, Section 1 of the Washington Constitution, the state asserted its sovereign ownership of the beds and shores of Puget Sound and other navigable tidal waters up to the line of ordinary high tide. The legislature, however, provided for the sale of tidelands—the lands between high and low tide—into private ownership between 1889 and 1971, when the sales were discontinued by statute. About 60 percent of the tidelands are now in private ownership (fig. 6.1). The waterward boundaries of privately owned tidelands may be either the line of *mean* low tide or the line of *extreme* low tide or the inner harbor line, depending on what law was in effect when the state originally sold the lands. These privately owned tidelands, however, are subject to federal, state, and local regulations.

Under current law, the state may sell tidelands only to public entities, but such lands may be leased to private parties (RCW 79.94.070, 79.94.150).

The Shoreline Management Act of 1971 (SMA)

The constitutionally mandated harbor line system and related statutes discussed above reserved public ownership and control over many commercially valuable harbors in Washington, but harbor lines were not

Figure 6.1 Signs along the shore alert the public to the private ownership of some tidelands.

established in all potential harbors and the system had its shortcomings. By the 1960s, industrial, commercial, and residential development was accelerating along the shores of Puget Sound and elsewhere (fig. 6.2). The Shoreline Management Act of 1971 (SMA) (RCW Chapter 90.58) was enacted in part as a response to concerns about this development and its effect upon the environment.

Among legislative findings cited in the SMA are that "much of the shorelines of the state and the uplands adjacent thereto are in private ownership . . . [and] that unrestricted construction on the . . . shorelines . . . is not in the best public interest." It is the state's policy, the SMA asserts, "to provide for the management of the shorelines . . . by planning for and fostering all reasonable and appropriate uses" (RCW 90.58.020).

The SMA, as an exercise of the state's police power, divides responsibility for managing land use along the shorelines between the state and local governments. The law provides for the zoning of a strip extending 200 feet landward from the ordinary high-water mark and requires permits before "substantial development," except for single-family residences and some other expressly exempted improvements, can take place in the regulated area.

In implementing the SMA, master programs were prepared by local governments—counties, cities, and towns—for regulation of shorelines consistent with guidelines adopted by the state Department of Ecology (DOE). Once these master programs had been approved by the de-

Figure 6.2 Typical high-density shorefront development in the Seattle metropolitan area.

partment, they constituted the primary land-use regulations along the shorelines.

The state guidelines provide that the local master programs are to contain plan elements covering economic development, public access, transportation, recreation, land uses on shorelines and adjacent areas, water uses, conservation, and historical/cultural aspects. The guidelines suggest that the framework for implementing shoreline policies and regulatory measures be provided by classifying shorelines into four distinct environments: natural, conservancy, rural, and urban (Washington Administrative Code [WAC], chapter 173-16).

While local governments had some latitude in tailoring their master programs to their particular jurisdictions, the state guidelines contain extensive definitions of the natural systems and criteria for various permissible uses. Moreover, compliance with the guidelines as adopted state regulations is required both in master program development and in permit application review.

The SMA prohibits any "development" on the shoreline "unless it is consistent with the policy of this [act] and . . . the applicable guidelines, rules, or master program" (RCW 90.58.140[1]). Permits from local governments are required for any "substantial development" (i.e., costing or valued at more than $2,500) within the areas covered by the SMA. However, the act exempts from permit requirements certain uses, including (1) normal maintenance or repair of existing structures; (2)

construction on wetlands of a single-family residence for an owner or lessee's own use if it is no higher than 35 feet above average grade level; (3) construction of noncommercial pleasure craft docks costing less than $2,500 for single-family residences; (4) construction of "the normal protective bulkhead common to single-family residences"; (5) "emergency construction necessary to protect property from damage by the elements"; (6) certain "construction and practices normal or necessary for farming, irrigation, and ranching activities"; and (7) "construction or modification of navigational aids" (RCW 90.58.030[3][e]).

"Substantial development"

When a permit for a "substantial development" is required under the SMA, the applicant applies to the appropriate local government (county, city, or town), which develops and exclusively administers the local permit system (RCW 90.58.140[3]). The SMA, however, sets forth a general procedure to be used in each local system, including publication of notice of application for permits, additional notification provisions such as posting of notices, and appeals of local governments' decisions on applications to the Shorelines Hearings Board or to the superior court (RCW 90.58.140[4],[5]).

The SMA also provides that "applicants for permits . . . have the burden of proving that a proposed substantial development is consistent with the criteria that must be met" before issuance of a permit, and that whoever requests review of the granting or denial of an application for a permit has the burden of proof (RCW 90.58.140[7]). In addition, the SMA authorizes the issuing authority to rescind a permit if, after a hearing following adequate notice, it finds that a permittee has not complied with permit conditions (RCW 90.58.140[8]).

The SMA's permit requirements are enforced by the state attorney general and the local governments' attorneys. The law provides that they "shall bring such injunctive, declaratory, or other actions as are necessary to insure that no uses are made of the shorelines . . . in conflict with the provisions and programs of [SMA] . . ." (RCW 90.58.210). Fines can be levied against persons "found to have wilfully engaged in activities on the shorelines . . . in violation of [SMA] or any of the master programs, rules, or regulations . . ." (RCW 90.58.220).

In addition to regulating shoreline land use through the master programs, the state Department of Ecology and local governments are authorized under the SMA to acquire lands and easements by purchase, lease, or gift (RCW 90.58.240).

Obtaining a shoreline permit

Most people become concerned when they learn that it's necessary to "obtain a permit" prior to developing, adding on, or physically constructing some improvements to their property. While the hassle of obtaining a permit is generally viewed as an obstacle, remember the permit is for your protection as well as your neighbor's. The permit requirements also protect you and your property from the activities of others whose actions can physically or economically injure your property. The message is: don't avoid getting the necessary permits before making some physical improvements to your property. In the long run, it is in the best interests of all to conform to the building and land-use codes of your local area.

When do I need a permit?

If your property is within 200 feet of the ordinary high-water mark (OHWM) and your development plans cost in excess of $2,500, you need a shoreline permit. Cities and counties around Puget Sound and Georgia Strait (as well as freshwater bodies) issue permits for what is defined as "substantial development," that is, projects costing more than $2,500. Some projects are exempt, for example, single-family residences, small docks, and bulkheads. However, while a shoreline permit may not be required, the projects must be constructed to meet the standards contained within the Shoreline Master Program. The shoreline permits are a way for the local jurisdiction to make sure that your plans, as well as those of your neighbors, are consistent with the locally adopted shoreline master plans. Some time back, each local government worked with its citizens and elected officials to develop a shoreline management plan. Each plan (called a Shoreline Master Program) was tailored by the city or county within policies and guidelines developed by the Washington State Department of Ecology. Most shoreline master programs include statements of goals and objectives about the use of the shoreline as well as use regulations. With only a very few exceptions, all of Washington's 39 counties and 160 incorporated towns have now adopted shoreline master programs—which include both saltwater and freshwater shorelines.

The state shoreline permit

An application for a shoreline permit to do some development will fall into one of the categories: (1) a *substantial development* permit—that is, any kind of development project costing more than \$2,500; (2) a *shoreline variance* permit—these are permits asking to "vary or differ" somewhat from the precise technical requirements of the master program regulations; and (3) a *shoreline conditional use* permit—these are permits to allow some use of the property different from what is specified in the shoreline master program, or some use that expressly calls the activity a "conditional use."

The application

The shoreline permit has three basic parts: (1) a brief statement of what is being planned, (2) a site plan, and (3) a vicinity map. Once a permit application has been submitted, the city or county must solicit public comments on your plans. The idea here is to see what the public, and especially neighbors, think of your plans. After all, you'd want to know if your neighbor was planning some substantial development! If you're smart, you'll discuss the plans with your neighbors ahead of time and work out any concerns or objections. If not, chances are that your neighbors, once alerted to your plans by local officials, will come to the public hearing prepared to object. The message here is, if possible, inform your neighbors ahead of time and iron out details or concerns before the public hearing. If your plan calls for work or alteration seaward of the OHWM, state and federal natural resource agencies will be interested. It is in your interest to contact them early. Your local planning agency can help.

State review

With the passage of the Shoreline Management Act in 1971, the State of Washington mandated the shorelines of the state to be special places. Because the shorelines are special places, all development activities must be approved not only at the local level but also reviewed by the State Department of Ecology. The review is a check by the state to be sure that the city or county is adhering to its shoreline plan. If the state believes the local jurisdiction erred in granting a substantial development permit, the state can appeal the permit to the state Shorelines Hearing Board. Such appeals are rare, but they do occur. For shoreline variance or conditional use permits, the DOE must approve or deny. If your permit is denied by the state, your recourse is to appeal to the Shorelines Hearing Board. This board hears all appeals, and any applicant or interested third party adversely affected by a decision of the state has the right of appeal. If you find yourself in this position, it is wise to hire a land-use attorney to act in your behalf.

The state permitting process: a word to the wise

The permit process for shoreline development requires approval at both local and state levels. Under "normal" conditions, permits are approved with little problem. Four things should be kept in mind when considering development:

1. Don't *avoid* applying for a permit.
2. *Talk* to your neighbors about your development plans.
3. The permit process takes *time*. Start your application months in advance of any construction dates.
4. Consider *consulting* an attorney specializing in land-use law to help you.

Federal permits and regulations

The coastal zone of the United States is also an area of interest to the federal government, and in two instances the federal government could affect your development plans. The U.S. Army Corps of Engineers is charged by Congress to oversee any activities within the "navigable waters" of the United States. Along the Pacific Coast, including Puget Sound and Georgia Strait, their jurisdiction extends to the mean higher high water mark (MHHW) along the shore. If your development plans extend seaward of this mark, you must seek a permit to do so from the Corps of Engineers. The trick is to locate where that line is on your shore. It is legally defined and can be located by a surveyor. Landward of the MHHW, no permit is needed by the Corps of Engineers. The kinds of activities normally requiring this permit are the installation of seawalls, bulkheads, riprap, or boat docks. The filling or draining of wetlands also requires federal permits.

In the event your development plans fall into an area of identified coastal flooding potential, as is often the case in Puget Sound, another federal agency becomes involved—the Federal Emergency Management Agency (FEMA). This agency provides federally subsidized flood insurance to individual homeowners in flood-prone locations. In order to qualify for this insurance, your city or county must have a floodplain management program, and you also must build your structure according to building codes that will help to minimize the effects of flooding (see below for more information).

Contact agencies

Local governments. To reach your city or county planning department or for information on building codes and codes administration, consult your phone directory under city or county government offices. Inquire whether or not you are required to seek a permit for your specific development plans.

State government. The Department of Ecology provides a centralized service to inform citizens what kind of permits might be required for specific projects, the agencies involved, and how to get in touch with them. The DOE's address is Washington State Department of Ecology, Environmental Permit Information Center, Mail Stop PV-11, Olympia, WA 98504.

Federal government. Addresses for the relevant federal agencies are U.S. Army Corps of Engineers, Seattle District, P.O. Box C-3755, Seattle, WA 98124, and Federal Emergency Management Agency, Region X, Federal Regional Center, 130 228th Street, S.W., Bothell, WA 98011.

The Washington State Coastal Zone Management Program (WSCZMP)

The Washington State Coastal Zone Management Program (WSCZMP) was the first such state program approved under the federal Coastal Zone Management Act of 1972 (CZMA) (P.L. 92-583), which is intended to encourage states to manage their shorelines and thereby conserve a vital national resource. The federal government approved Washington's original program in 1976 and amendments to it in 1979.

At the heart of the WSCZMP is the comprehensive control program instituted under the previously discussed Shoreline Management Act of 1971 (SMA). While the SMA planning process and regulatory permit system are initiated at the local level, the state DOE has played a key role, preparing and adopting statewide guidelines that must be complied with by local governments.

In implementing the coastal management program, the SMA is buttressed by two state laws, the State Environmental Policy Act of 1971 (SEPA) (RCW Chapter 43.21C) and the Environmental Coordination Procedures Act of 1973 (ECPA) (RCW Chapter 90.62).

SEPA is a comprehensive environmental law containing both substantive policies and procedural directives. All state and local government authorizations of any projects must comply with SEPA by considering environmental and ecological factors when taking "major actions significantly affecting the quality of the environment" (RCW 43.21C.030).

Figure 6.3 The ecological value of large wetland and tideland sites has encouraged the public to designate these areas as preserves. Padilla Bay in Skagit County is a National Estuarine Sanctuary.

Among environmental policies enunciated in SEPA is the declaration that "each person has a fundamental and inalienable right to a healthful environment and that each person has a responsibility to contribute to the preservation and enhancement of the environment" (RCW 43.21C.020[3]). SEPA requires the preparation of environmental impact statements (EISs) for many types of proposed development along the Puget Sound area's shoreline.

The Environmental Coordination Procedures Act of 1973 (ECPA) establishes a coordinated system for persons applying for various state and local permits, including those required under the SMA master programs. Master applications may be submitted to the state Department of Ecology or to designated local governmental offices, and the department notifies the applicant of the requirements of all agencies covered by ECPA.

The DOE, which is in charge of many of the environmental regulations in the state, is the lead agency administering the coastal management program. The department is the basic authority for enforcing such regulatory programs as those concerning water resources, water quality, air quality, and air pollution. Among other state agencies in the coastal management network are the Department of Natural Resources, the Shorelines Hearings Board, Pollution Control Hearings Board, Parks and Recreation Commission, Department of Fisheries, and Department of Game.

The coastal management program identifies various "areas of particular concern" (fig. 6.3). Selection was based on these criteria: "(1) the area contains a resource feature of environmental values considered to be of greater than local concern or significance; (2) the area is given recognition as of particular concern by state or federal legislation, administrative and regulatory programs, or land ownership; and (3) the area has the potential for more than one major land or water use or has a resource being sought by ostensibly incompatible users."

Among the areas of particular concern are (1) the Nisqually Estuary, (2) Hood Canal, (3) Snohomish River Estuary, (4) Skagit Bay and Padilla Bay, (5) the Northern Strait and Puget Sound Petroleum Transfer and Processing Area, and (6) the Dungeness Estuary and Spit Complex. Many of these areas parallel the "shorelines of statewide significance" listed in the Shorelines Management Act (RCW 90.58.030[2][e]). Under the SMA, the state, through the Department of Ecology, has greater planning authority over this category of shorelines than for ordinary shorelines.

Miscellaneous state and local laws

A private residential owner of property adjoining state-owned shorelands or tidelands, except for those in harbor areas, has a qualified right to install and maintain without charge a private recreational dock under state law (RCW 79.90.105). This privilege is subject to local regulation governing the dock's construction, size, and length. The state Department of Natural Resources may revoke the permission in certain circumstances.

In saltwater harbor areas, the state may issue leases for wharves, docks, and similar structures (RCW 79.92.060). The Department of Natural Resources administers the leasing program (fig. 6.4). If the department decides to lease state-owned tidelands and shorelands of the first class (within or near cities), the owner of the adjoining uplands has a preference to apply to lease the lands (RCW 79.94.070).

In addition to fulfilling the permit requirements resulting from the Shoreline Management Act, persons constructing improvements along the shoreline must comply with relevant local building codes and other ordinances (RCW 90.58.360). Counties and municipal corporations have the authority to enact building codes and to regulate structures (RCW 35.22.280, 35.23.440, 35.24.290, 36.43.010).

County commissioners are charged with flood prevention and are authorized to make necessary improvements (RCW 36.32.280, 86.12.020).

Figure 6.4   At Cherry Point in Whatcom County the State of Washington is trying to balance the needs for development with the mandate to protect the marine environment.

National Flood Insurance Program (NFIP)

The National Flood Insurance Act of 1968 (P.L. 90–448) as amended by the Flood Disaster Protection Act of 1973 (P.L. 93–234) was passed to encourage prudent land-use planning and to minimize property damage in flood-prone areas, including the coastal zone. Local communities must adopt ordinances to reduce future flood risks to qualify for the National Flood Insurance Program. The NFIP provides an opportunity for property owners to purchase flood insurance that generally is not available from private insurance companies.

The initiative for qualifying for the program rests with the community, which must get in touch with the Federal Emergency Management Agency (FEMA). Any community may join the National Flood Insurance Program provided that it requires development permits for all proposed construction and other development within the flood zone and ensures that construction materials and techniques are used to minimize potential flood damage. At this point the community is in the "Emergency Phase" of the NFIP. The federal government makes a limited amount of flood insurance coverage available, charging subsidized premium rates for all existing structures and/or their contents, regardless of the flood risk.

FEMA may provide a more detailed Flood Insurance Rate Map (FIRM) indicating flood elevations and flood-hazard zones, including velocity zones (V-zones) for coastal areas where wave action is an additional hazard during flooding. The FIRM identifies Base Flood Elevations (BFES), establishes special flood-hazard zones, and provides a basis for floodplain management and establishing insurance rates.

To enter the Regular Program phase of the NFIP, the community must adopt and enforce floodplain management ordinances that meet at least the minimum requirements for flood-hazard reduction as set by FEMA. The advantage of entering the Regular Program is that increased insurance coverage is made available, and new development will be more hazard-resistant. All new structures will be rated on an actual risk (actuarial) basis, which may mean higher insurance rates in coastal high-hazard areas, but generally results in a savings for development within numbered A-zones (areas flooded in a 100-year coastal flood but less subject to turbulent wave action).

FEMA maps commonly use the "100-year flood" as the base flood elevation to establish regulatory requirements. Persons unfamiliar with hydrologic data sometimes mistakenly take the "100-year flood" to mean a flood that occurs once every 100 years. In fact, a flood of this magnitude could occur in successive years, or twice in one year, and so on. If we think of a 100-year flood as a level of flooding having a 1 percent statistical probability of occurring in any given year, then during the life of a house within this zone that has a 30-year mortgage, there is a 26 percent probability that the property will be flooded. The chances of losing your property become 1 in 4, rather than 1 in 100. Having flood insurance makes good sense.

In V-zones new structures will be evaluated on their potential to withstand the impact of wave action, a risk factor over and above the flood elevation. Elevation requirements are adjusted, usually 3 to 6 feet above still-water flood levels, for structures in V-zones to minimize wave damage, and the insurance rates are also higher. When your insurance agent submits an application for a building within a V-zone, an elevation certificate that verifies the elevation of the first floor of the building must accompany the application.

The insurance rate structure provides incentives of lower rates if buildings are elevated above the minimum federal requirements. General eligibility requirements vary among pole houses, mobile homes, and condominiums. Flood insurance coverage is provided for structural damage as well as contents. Most coastal communities are now covered under the Regular Program. To determine if your community is in the NFIP and for additional information on the insurance, get in touch with your local property agent or call the NFIP's servicing contractor (phone: [800] 638-6620), or the NFIP Region X Office at the Federal Regional

Center, 130 228th Street, S.W., Bothell, Washington 98011 (phone: [206] 481-8800). For more information, request a copy of "Questions and Answers on the National Flood Insurance Program" from FEMA.

Before buying or building a structure on the coast, an individual should ask certain basic questions:

Is the community I'm located in covered by the Emergency or Regular Phase of the National Flood Insurance Program?

Is my building site above the 100-year flood level? Is the site located in a V-zone? V-zones are high-hazard areas and pose serious problems.

What are the minimum elevation and structural requirements for my building?

What are the limits of coverage?

Make sure your community is enforcing the ordinance requiring minimum construction standards and elevations. Most lending institutions and local governmental planning, zoning, and building departments will be aware of the flood insurance regulations and can provide assistance. It would be wise to confirm such information with appropriate insurance representatives. Any authorized insurance agent can write and submit a National Flood Insurance policy application. All insurance companies charge the same rates for National Flood Insurance policies.

Disaster Relief Act

The FEMA Disaster Assistance Program Division serves as an advisory agency for the reduction of impacts due to natural hazards (for example, flooding, landslides, earthquakes), as well as exerting some regulatory control to reduce future property damage. Under the authority of the federal Disaster Relief Act of 1974 (P.L. 93–288) the agency evaluates potential hazards and determines plans to mitigate the effects of such hazards. Reduction of loss due to flooding is specifically addressed under the authority of the Inter-Agency Agreement for Nonstructural Flood Damage Reduction and Executive Order 11–988 Flood Plain Management, which designates FEMA as the lead agency to determine actions that will reduce the impact of flooding. For more information, call FEMA, Region X, Disaster Assistance Program Division, (phone: [206] 481-8800).

Other federal programs

Several other federal programs are applicable to the Puget Sound area's coastal zone. Coastal residents, property owners, officials, and developers should be aware of these programs.

The previously discussed Washington State Coastal Zone Management Plan was developed under and complies with the federal Coastal Zone Management Act of 1972 (P.L. 92–583). The act is administered by the Office of Coastal Resources Management, National Oceanic and Atmospheric Administration, Department of Commerce.

The U. S. Army Corps of Engineers is the federal agency responsible for regulating dredging and filling in the Puget Sound area. It is charged with administering the Rivers and Harbors Act of 1889 (33 USC 403) and the Federal Water Pollution Control Act of 1972 (P.L. 92–500), as amended.

Other federal laws with a major impact on the area's coastal zone are administered by the Environmental Protection Agency, the Fish and Wildlife Service, and the National Marine Fisheries Service.

In addition to the laws noted above, other federal regulations may be important locally. Coastal residents should check with the state Department of Ecology or local governmental planning, zoning, and building departments.

Summary

During the past 15 years, land-use planning and regulation in the Puget Sound area's coastal zone has increased dramatically. The Shoreline Management Act of 1971 (SMA), which anticipated the federal Coastal Zone Management Act (CZMA) passed the following year, established a framework for creation of locally initiated master programs that must be consistent with statewide guidelines. Under these programs, a 200-foot-wide strip around the sound is zoned, and permits must be obtained for many types of "substantial development."

The SMA is the cornerstone of the Washington State Coastal Zone Management Program, the first such state program to be approved by the federal government under the CZMA. The National Flood Insurance Program and the U.S. Army Corps of Engineers' dredging and filling permit system are among important federal programs having an impact on coastal land use.

With the continuing population growth in the Puget Sound area, coupled with concerns for the environment, additional land-use planning and regulation may be expected in the future.

If this book has in some small way informed the reader or caused a change in attitude or action, it has been successful. The preceding pages are not aimed at frightening coastal residents or stifling development; instead, they are meant to arm users of the coastal zone with insights about the dynamics of this ever-changing environment. Development of coastal Puget Sound and Georgia Strait will continue, as it should; however, it must proceed with a recognition of and appreciation for the coastal processes of the region.

Readers might justifiably question the risk classifications given to particular sections of coast. Please keep in mind that these designations are both general and subjective. A prudent approach was taken when assigning the classifications. Labels of high risk were fixed liberally. Years of coastal study have shown that change tends to occur episodically. Years may pass with little erosion, landsliding, or flooding. Then that singular "rare" storm event strikes, forever changing the landscape and wiping out a lifetime investment.

When considering a particular coastal site for investment, be cautious, be prudent. Carefully inspect the site. Try to interpret what the coastal landscape might be revealing to you. Should you have some unanswered questions or concerns about the physical characteristics of a particular site, seek the assistance of a professional coastal geologist, geographer, or engineer. Remember that the placid conditions of summer can dramatically change during the raging storms of winter. Finally, consider a little extra precaution when developing a coastal property. The absolute best advice is to build your home and other structures an added measure back from the sea cliff or water's edge.

Puget Sound and Georgia Strait have evolved into the unique beauty we all enjoy today. Those evolutionary processes are still operating, continually molding and shaping the shoreline. Our goal should not be to challenge the physical evolution of this coast but to try and live in harmony with it.

Appendix

Useful references

Beaches and beach processes

Washington Public Shore Guide: Marine Waters, by T. W. Scott and M. A. Revling, 1986. Seattle: University of Washington Press. A detailed, comprehensive, county-by-county guide to public facilities along Washington's shore. A must for any family that goes to the beach.

Beach Processes and Sedimentation, by P. D. Komar, 1976. Englewood Cliffs, N.J.: Prentice-Hall. A thorough but technical presentation of beach processes and coastal sedimentation. Most useful to engineers and those with a solid math and/or scientific background.

Our Changing Coastlines, by F. P. Shepard and H. R. Wanless, 1971. New York: McGraw-Hill. This book provides a descriptive overview of historical changes to our nation's coastline. It is very well illustrated and informative. Best suited to those readers seeking some general information about regional coastal change.

The Beaches Are Moving, by Wallace Kaufman and Orrin Pilkey, Jr., 1984. Durham, N.C.: Duke University Press. A very readable and informative presentation of coastal problems and prospects around the country. This book is a logical next step to the *Puget Sound* book.

Waves and Beaches, by Willard Bascom, 1964, 1982. Garden City, N.Y.: Doubleday. This book is a well-known classic in the study of waves and beaches. It is both informative and entertaining and is aimed at a general readership.

The World's Coastline, by E. C. F. Bird and M. L. Schwartz, 1985. New York: Van Nostrand Reinhold. This recently completed book compiles the contributions of many coastal researchers around the world, describing the physical characteristics of their respective countries' shores.

The Coast of Puget Sound: Its Processes and Development, by John

Downing, 1983. Seattle: University of Washington Press. Downing's book is very well illustrated and written, focusing on coastal processes within Puget Sound. Highly recommended for those readers wanting more in-depth information about this coastal region.

Coastal Zone Atlas of Washington (15 vols.), 1980. State of Washington, Department of Ecology, Shorelands Division. The Department of Ecology has prepared a coastal atlas for each of the counties bordering Puget Sound, Georgia Strait, and the Strait of Juan de Fuca. These maps are extremely useful references for site-specific information on coastal geology, slope stability, land use, flooding potential, and critical biological areas. Local libraries usually have the atlas for their respective county. The atlases can be purchased directly from the State of Washington, Department of Ecology, Shorelands Division, PV-11, Olympia, WA 98504.

Shoreline erosion and protection

Shore Protection Manual, 4th ed. (vols. 1–2), U.S. Army Corps of Engineers, 1984. Washington, D.C.: U.S. Government Printing Office, Publication #008-022-00218-9. A very technical manual on coastal processes and engineering.

Low Cost Shore Protection: A Guide for Local Government Officials, U.S. Army Corps of Engineers, 1981.

Low Cost Shore Protection: A Property Owner's Guide, U.S. Army Corps of Engineers, 1981.

Low Cost Shore Protection: A Guide for Engineers and Contractors, U. S. Army Corps of Engineers, 1981. The government has produced the three very good resource documents listed above for the coastal property owner, local government officials, and engineers. They are well illustrated and generally informative. Copies are available from the U.S. Army Corps of Engineers, Seattle District, P.O. Box C-3755, Seattle, WA 98124, or from U.S. Army Corps of Engineers, district or division offices in your region.

Shore Erosion Protection: On Shore Structural Methods, by Douglas Canning, 1985. State of Washington, Department of Ecology, Shorelands Division.

Coastal slope stability

Landslides in Seattle, by Donald W. Tubbs, 1974. State of Washington, Department of Natural Resources, Division of Geology and Earth

Resources, Information Circular No. 52. A valuable document for those concerned about the condition and location of landsliding in Seattle. Available in local libraries or directly from the Washington State Department of Natural Resources, Division of Geology and Earth Resources, Olympia, WA 98504.

Slope Stability Map of Thurston County, Washington, by Ernest R. Artrim, 1976. State of Washington, Department of Natural Resources, Division of Geology and Earth Resources, Geologic Map GM-15.

Relative Slope Stability of the Southern Hood Canal Area, Washington, by Mackey Smith and R. J. Causen, 1977. U.S. Geological Survey Miscellaneous Investigations Series Map I-853-F.

Relative Slope Stability of Gig Harbor Peninsula, Pierce County, Washington, by Mackey Smith, 1976. State of Washington, Department of Natural Resources, Division of Geology and Earth Resources, Geologic Map GM-18.

Shoreline Bluff and Slope Stability: Technical Management Options, by Douglas Canning, 1985. State of Washington, Department of Ecology, Shorelands Division. Another of the useful information papers prepared by the Division of Ecology and aimed at informing residents about specific land development and slope-hazard problems. Available from the State of Washington, Department of Ecology, Shorelands Division, PV-11, Olympia, WA 98504.

Engineering Geologic Studies, 1976. State of Washington, Department of Natural Resources, Division of Geology and Earth Resources, Information Circular No. 58.

Coastal flooding

Flood Insurance Study, U.S. Department of Housing and Urban Development, Federal Insurance Administration. Technical reports containing maps and flood data on a county-by-county basis. Available from local libraries or from the Federal Emergency Management Agency, Region X, Federal Regional Center, 130 228th Street, S.W., Bothell, WA 98011.

Flood Hazard Management Profiles, 1984. Center for Urban and Regional Studies, University of North Carolina, Chapel Hill, N.C. 27514.

Preparing for Hurricanes and Coastal Flooding: A Handbook for Local Officials, 1983. Federal Emergency Management Agency, Publication #50, P.O. Box 8181, Washington, D.C.

Flood Emergency and Residential Repair Handbook, 1986. Federal Emergency Management Agency, Publication #FIA-13, Washington, D.C.

Floodproofing Non-Residential Structures, 1986. Federal Emergency Management Agency, Publication #FEMA 102, Washington, D.C.

Manufactured Home Installation in Flood Hazard Areas, 1985. Federal Emergency Management Agency, Publication #FEMA 85, Washington, D.C.

Coastal vegetation

Erosion and Sedimentation Control Manual, by Tony Barrett, 1982. State of Washington, Department of Ecology, WDOE Dept. 82-3, Olympia, WA.

Designing for Bank Erosion Control with Vegetation, by P. L. Knutson, 1978. U.S. Army Coastal Engineering Research Center, P.O. Box 631, Vicksburg, MS 39180.

Planting Guidelines for Marsh Development and Bank Stabilization, by P. L. Knutson, 1977. U.S. Army Coastal Engineering Research Center, Box 631, Vicksburg, MS 39180.

The Role of Vegetation in Shoreline Management. Great Lakes Basin Commission, P.O. Box 999, Ann Arbor, MI 48106.

Coastal construction and site analysis

Coastal Design: A Guide for Planners, Developers, and Homeowners, by Orrin H. Pilkey, Sr., Walter D. Pilkey, Orrin H. Pilkey, Jr., and William J. Neal, 1983. New York: Van Nostrand Reinhold.

Design and Construction Manual for Residential Buildings in Coastal High Hazard Areas, 1980. Washington, D.C.: U.S. Department of Housing and Urban Development and Federal Emergency Management Agency.

Coastal conservation and planning

Coastal Ecosystems: Ecological Considerations for Management of the Coastal Zone, by John Clark, 1972. Washington, D.C.: The Conservation Foundation. A clearly written, well-illustrated book on the applications of principles of ecology to the major coastal zone environments.

Design with Nature, by Ian McHarg, 1971. Garden City, N.Y.: Doubleday. A classic text on designing with the environment. Stresses man's need to develop the land recognizing nature and ecological processes.

The Water's Edge: Critical Problems of the Coastal Zone, by Bostwick H. Ketchum (ed.), 1972. Cambridge, Mass.: MIT Press. A good but somewhat dated presentation of a broad spectrum of coastal issues.

Managing Washington's Shores: Today's Challenge, 1985. State of Washington, Department of Ecology, Shorelands Division. A small booklet describing shoreline management legislative mandate concerns. Available by mail request, State of Washington, Department of Ecology, Shorelands Division, PV-11, Olympia, WA 98504.

Coastal permits and regulations

Permit Program: A Guide for Applicants. Washington, D.C.: U.S. Army Corps of Engineers (EP 1145-2-1).

Regulations to Reduce Coastal Erosion. Wisconsin Coastal Zone Management Program, State Office of Planning and Energy, GEF II, 101 S. Webster St., Madison, WI 53702.

Applications for Department of the Army Permits for Activities in Waterways, 1974. Washington, D.C.: U.S. Army Corps of Engineers, Office of Chief of Engineers. Booklet describing coastal activities and structures requiring permits from the U.S. Army Corps of Engineers. Available by writing U.S. Army Corps of Engineers, Seattle District, P.O. Box C-3755, Seattle, WA 98124.

Index

About the Authors

Thomas A. Terich is a member of the faculty of the Department of Geography and Regional Planning at Western Washington University, Bellingham. He was formerly a city planner in southern California where he developed an interest in coastal management. He received a Ph.D. in geography at Oregon State University in 1973 and has maintained a research specialization in shore processes and coastal management.

Peter H. F. Graber is an attorney specializing in land title and boundary disputes, with an emphasis on tide and submerged land litigation. His office is located in Marin County, California. He formerly was Deputy Attorney General, Office of the Attorney General, State of California. He has practiced law since 1962. He received a B.A. in political science and an M.S. in journalism from UCLA, and his law degree from the University of California, Berkeley (Boalt Hall).

Library of Congress Cataloging-in-Publication Data
Terich, Thomas.
Living with the shore of Puget Sound and the Georgia Strait.
(Living with the shore)
"Sponsored by the National Audubon Society."
Bibliography: p.
Includes index.
1. Shore protection—Washington (State)—Puget Sound
Region.   2. Coastal zone management—Washington (State)—
Puget Sound Region.   3. Shore-lines—Washington (State)—
Puget Sound Region.   4. Beach erosion—Washington (State)
—Puget Sound Region.   5. House construction—Washington
(State)—Puget Sound Region.   I. National Audubon Society.
II. Title.   III. Series.
TC224.W2T47   1987   333.91'7'097 977   86-29174
ISBN 0-8223-0689-1
ISBN 0-8223-0745-6 (pbk.)